Disclaimer

The publisher of this book is by no way associated with the National Institute of Standards and Technology (NIST). The NIST did not publish this book. It was published by 50 page publications under the public domain license.

50 Page Publications.

Book Title: Correlation Filters for Elastic-Distorted Live-Scan Fingerprint Recognition

Book Author: D P. Casasent; Craig I. Watson

Book Abstract: This is a summary of a multi-year study of the use of distortion-invariant filters for recognition of live-scan dab fingerprints with elastic distortions. These fingerprints are characterized by elastic distortions. NIST special database 24 is used; it represents the only available database containing a number of elastic distortions for each fingerprint. Several different distortion-invariant filters were addressed including two new ones that used high-pass filtered fingerprint data to improve discrimination. Verification and identification applications are addressed and test procedures for evaluating algorithms/systems for each are defined.

Citation: NIST Interagency/Internal Report (NISTIR) - 6990

Keyword: biometric recognition;distortion-invariant filters;elastic distortions;fingerprint recognition

NISTIR 6990

Correlation Filters for Elastic–Distorted Live-Scan Fingerprint Recognition

Dave Casasent
Craig Watson

National Institute of Standards and Technology
Technology Administration, U.S. Department of Commerce

NISTIR 6990

Correlation Filters for Elastic–Distorted Live-Scan Fingerprint Recognition

Dave Casasent
Carnegie Mellon University

Craig Watson
Information Technology Laboratory
Information Access Division

June 2003

U.S. DEPARTMENT OF COMMERCE
Carlos M. Gutierrez, Secretary
TECHNOLOGY ADMINISTRATION
Michelle O'Neill, Acting Under Secretary of Commerce for Technology
NATIONAL INSTITUTE OF STANDARDS AND TECHNOLOGY
William A. Jeffrey, Director

Specific hardware and software products identified in this report were used to conduct work described in this document. In no case does identification of any commercial product, trade name, or vendor imply recommendation or endorsement by NIST, nor does it imply that the products and/or equipment identified are necessarily the best available for the purpose.

Correlation Filters for Elastic-Distorted Live-Scan Fingerprint Recognition

David Casasent (casasent@ece.cmu.edu) and
Craig Watson (craig@magi.nist.gov)

A. EXECUTIVE SUMMARY

This is a summary of a multiyear study of the use of distortion-invariant filters for recognition of live-scan dab fingerprints (FPs) with elastic distortions. These FPs are characterized by elastic distortions. NIST special database 24 is used; it represents the only available database containing a number of elastic distortions for each FP. Verification and identification applications are addressed and test procedures for evaluating algorithms/systems for each are defined for the first time.

A training set of 10 sec of live-scan dab FP data are used as the training set from which a single distortion-invariant filter for each FP is formed. For 55 FPs, at least 9 elastic distorted FPs were available, eight were used to synthesize the filters and one was used for testing. This is a quite large set of FPs with multiple elastic distortions per FP, compared to any prior data; that was the purpose of producing database 24. These 55 FP sets were extensively tested; the other 145 FP sets were tested less fully.

The emphasis was whether a filter formed from eight distorted FPs was sufficient to capture and recognize the elastic distortions to be expected. Such a training set can easily be acquired in a very short training session. Elastic distortions were the major FP variation intended to be addressed. However, many of the FPs were also dry, partial, oily or scarred. Four sets of FP data were produced using coarse and fine-aligned data (rotation and shift alignments were considered) and using unnormalized and energy normalized data.

Use of normalized data was found to improve results for dry and oily FPs and is expected to improve results for partial FPs. As expected, use of fine-aligned data performed better or comparable to use of coarse-aligned data.

Several different distortion-invariant filters were addressed including two new ones that used high-pass filtered FP data to improve discrimination. The Minace filter was found to perform best. It gave perfect results for verification (if normalized and fine-aligned data were used). For identification, Minace filters with all four types of data gave perfect results.

These filters seem to perform better than standard minutiae matching methods on FPs of poor quality (dry, oily, partial) that would be rejected. Thus, such filters clearly merit further attention.

ABSTRACT

Sets of live-scan dab data for a large database of fingerprints (FPs) with elastic distortions are used to evaluate the use of distortion-invariant filters for FP recognition. Test procedures for FP verification and identification are advanced for the first time. One filter is formed for each FP. A

set of 55 FPs are extensively analyzed. We find that a training set of eight FPs with elastic distortions are sufficient to capture these variations. For verification applications, filters formed from fine-aligned and normalized distortion-invariant data are found to produce perfect test results. For identification applications, filters formed form coarse-aligned and unnormalized data produce perfect test results. Oily, dry, partial and scarred FPs are all handled. Standard minutiae matching methods would reject many FPs handled by these distortion-invariant filters.

1. INTRODUCTION

NIST (The National Institute for Standards and Technology) has considered an optical FPC (frequency plane correlator) interfaced to an optical memory [1] as a system that allows it to address if optical correlation of an input versus a large number of reference patterns is of use. A digital correlation realization is also under current consideration. One application considered is fingerprint (FP) recognition. We consider the case when the FP input is obtained by the user dabbing his finger onto a live scan FP reader (the user touches his print onto the reader once). Such FP images differ considerably from rolled FP images, since they contain a wider range of elastic distortions due to variations in the amount and distribution of pressure applied across the FP when it is entered into the FP reader. These elastic distortions are much more varied than those present in rolled prints. FP recognition systems must operate in the presence of such distortions and in the presence of rotations of the FP. Our present work emphasizes FP recognition in the presence of elastic distortions using distortion-invariant correlation filters. No prior correlation filter FP recognition work has fully addressed elastic distortions. We remove/reduce rotational differences between FPs by preprocessing to allow us to address the effects of mainly elastic distortions. We would employ similar preprocessing methods in a final system (otherwise, rotated images of each FP must be included in the training set for each distortion-invariant filter) and the number of training images needed would be quite large). All correlation peak output data reported are the magnitude of the correlation peak.

The NIST optical correlation laboratory demonstration system [1] used a Kopin LC (liquid crystal) to input a number of versions of a user's FP, these were all summed and a simple averaged composite filter was formed on a thermoplastic (TP) device as the reference filter. The user then inputs a new version of his FP to the LC, it was correlated with the filter on the TP. In over 50 real-time demonstrations to various visiting groups, this always worked; the user was identified, and other input prints were not. Issues such as dirt, etc. on the TP, the partial frame integration of LC data (due to LC rise and fall times), etc. are not considered in the present advanced version of such filters. *It would be attractive to fully model and digitally simulate this system, but this is beyond the scope of present effort levels*. More details on this optical laboratory system are available [in Appendix A1]. Its success was the motivation for this present effort. Our present effort concerns the use of advanced distortion-invariant filter methods to form such filters from training set images.

The FP problem considered is a limited access one (a database of ≈1000 FPs). We will make one filter for each person in the database. In the assumed scenario, each authorized person has submitted a number of live scan FP dab images of his FP; these are assumed to be representative of the elastic distortions expected. A separate distortion-invariant filter is formed for each FP from its training set images and is then used to recognize a different live scan input of the FP for verification and/or identification. Use of a hierarchical set of filters is also possible, but was not considered. There are many FP databases available from NIST (see www.itl.nist.gov/div894/894.03). However, they are all of rolled FPs and contain only one or two versions of each FP. There is no known available database of live scan dab FP images that provide the range of expected elastic distortions. The Michigan State University (MSU) FP database consists of 700 live scan FP images, with ten images per finger. About 10% of the images are not of good quality. Many of the images of the same FP have only a small common FP region [29]. This occurred since no restrictions were placed on the position of the fingers during capture. Thus, a new elastic distorted FP database was produced for this study; it is publicly available as NIST Special Database 24. No prior published work has addressed the effect of elastic distortions on various FP recognition systems. NIST is in the process of formulating test and evaluation procedures for various biometric identification systems (most commercial FP systems use minutia matching). In these systems, the threshold is set for a given P_{FA} (probability of a false alarm, recognition of an incorrect FP as a given true FP; it is the percentage of all possible false alarms), a typical desired P_{FA} is 0.001 (i.e. 0.1% errors or one in 1000). When the threshold for one commercial system was set for this P_{FA}, it gave a rejection rate $P_R = 0.15$, i.e. 15% of the input prints were not accepted (no decision was made on them). For these initial FPs, the user would be asked to reenter his FP. This P_R rate is too high and should be reduced. These tests were made on a different database from the present one. The correct recognition rate P_C is the percentage of all FPs in the test set database that were correctly recognized. We consider verification (the user enters his FP and some indicator of his identity; the correct filter must give an output above some threshold) and identification (the user only enters his FP and any filter giving an output above some threshold will admit the user).

Prior Work (Digital)

Most FP recognition efforts use digital techniques. They can involve extensive preprocessing to enhance the FP [2,3]. Most FP recognition systems devote extensive processing to noise removal, image enhancement, etc. [4,28,29] before and after minutiae extraction and before and after binarization of the FP image. Our work did not employ any such techniques, future work could utilize such methods for our training set and for the test set input also. Most FP recognition systems also require global attributes such as location of the core (the center of the FP) and delta points (where ridge flow splits) and finally minutiae points (ridge endings or bifurcations) [4,29]; this can take excessive computing times (2 sec for processing and minutae extraction [M. McCabe (Pers. Corresp. 8/01]). Once minutiae points have been located, the FP matching is fast, even when the minutiae points are shifted due to elastic distortions [4,29] (40s to search 1.75×10^6 prints or 175K cards; [M. McCabe, Pers. Corresp. 8/01]). All FP systems using minutiae matching require high-quality and high-resolution FP images, since without high quality FP imagery minutiae cannot be reliably located. They do not perform well on FPs with scars and they generally have a high reject rate P_R (the number of FPs on which no decision is made). In digital FP recognition systems using minutiae points, various approaches [29] are used to accommodate efficient matching of these points when elastic distortions are present. At least 25 minutiae must often [29] be located to achieve high performance. FPs in which this number of minutiae cannot easily be located are generally rejected by present systems, which is another reason why present minutiae-based FP systems are characterized by high P_R rates.

Most FP recognition systems use binarized images [28], a summary of these binarization techniques exists [28]. Our approach to binarization is very simple, compared to others.

NIST [15] has conducted extensive FP recognition tests using neural networks (NNs); however rolled prints rather than live scan ones were used, and elastic distortions were not considered. Each FP was coarsely centered using the NIST core location algorithm. Each FP, was partitioned into $4 \times 4 = 16$ square regions centered on the core of the FP. These 16 regions of a test FP were correlated (with shifts) with the 16 image regions of a reference FP. This was repeated for a second set of 16 regions with the core shifted, to allow for errors in the core location algorithm. The 32 correlation peak values were fed to a NN, whose output is 1 for a match and 0 otherwise. The NN weights were trained on 1000 pairs of prints (2 per finger); the 32 correlation values for the 1000 matching fingers were fed to the NN to provide an output of 1; for all other non matching pairs of prints, their 32 correlation values were fed to the NN to produce an output of 0. With the NN trained and the weights fixed, it was tested on another separate set of 1000 FP pairs. Shifted FP subimages were considered to handle core location errors and rotational differences were removed; $P_C = 84.3\%$ was obtained.

Prior Work (Optical)

Optical processing for FP recognition has generally used an optical correlator. Recent device advances make gray-scale optical correlators with 512×512 pixel inputs and filters possible with 1000 correlation/sec data rates (without using phase-only filters) [13,14]. Thus, optical correlator approaches are attractive because of their speed. However, they use global ridge and valley FP data, rather than FP minutiae and thus are susceptible to larger P_{FA} (false alarm probability) rates than are minutiae-based methods unless other methods are considered. The most obvious technique to improve P_C is [5] to use M multiple samples of an input FP; in this case, assuming independent samples, $P_{RM} = (P_R)^M$ and $P_{FAM} = 1 - (1-P_{FA})^M$, where P_R and P_{FA} are for one test sample and P_{RM} and P_{FAM} are for M test samples of the same FP. This reduces P_R (exponentially), but it increases P_{FA} (linearly).

The first commercial FP optical correlation verification system uses this approach [5] with composite filters [6] different from those we consider (they optimize different correlation plane parameters). In training, each person slides his finger through a scanner producing many images, four images are selected and a filter is formed from them. In verification tests, the user repeats the process for $\approx$4-10 sec; all FP images are optically analyzed vs. the expected filter on-line at 60 images/sec. Various correlation plane parameters are used, we use only correlation peak height in our initial work. They achieve very good $P_R = 0.9\%$ and $P_{FA} = 0.2\%$ rates, but only 20 FPs were used and multiple test inputs per FP were employed. In our approach, more FPs are used (55); we use at least eight images per FP to form our filters (vs. four images in prior work [5]); we use only one test input for each FP; we use different filters and preprocessing (to handle shifts and rotations); and we emphasize elastic distortions due to dab images (sliding an FP reduces elastic distortions, but produces smear and other types of distortions).

There are at least three issues to consider in optical correlation FP recognition: rotation effects (how accurately must the test FP be aligned), distortion effects (especially elastic distortions), and image preprocessing. Other obvious issues are: use of a large database, FP resolution, etc. In the available literature, we have not seen a systematic approach to analyze the use of correlation filters for FP recognition; different researchers use different databases.

Our major present concern is whether correlation filters are of use for FP recognition when elastic distortions are present. There is limited attention to this issue in prior work (due to lack of a database). The Mytec Technologies Inc. results [5,6] are the best, but they considered only 20 FPs, not elastic distortions (produced by dab FP inputs), and used multiple observations for each test FP. In one test [7] on 20 elastic distorted versions of six FPs, P_C was poor (80%). In another test [8] using only six distorted inputs, $P_C = 80\%$ was also obtained. These last two tests did not use distortion-invariant filters and the databases were very small. Published results on large databases are limited; in one case, 3500 FPs of the same class/type were used, but results on six distorted test set versions of each FP gave poor $P_C = 83\%$ results (again, without use of distortion-invariant filters that we consider). Our present work includes the formation of a new publicly-available large database of a number of elastic distorted FP images of 100 (or all 200 available?) different FPs.

Many optical FP recognition efforts have addressed the rotational sensitivity of optical correlators (by how much does the correlation peak drop due to a misalignment θ between the test input and the reference). Different work [7,9,24,25,26] reports that $\pm 2°$ to $\pm 3°$ rotational differences result in a 50% drop in the correlation peak value (for auto-correlations). However, the FP image resolution varied in these tests; lower resolution images are expected to provide better rotational tolerances, but poorer P_C and P_{FA} results. Prior work has not addressed this issue of FP resolution effects. In our work, we propose to digitally align test inputs to a reference orientation; this allows emphasis primarily on elastic distortions. We also address digital rotation effects on correlation peak loss that have not been largely considered in prior work.

The preprocessing applied to an FP test (or training set) input are also of concern. In an all-optical system, such preprocessing is difficult, but possible, when the optical architecture is a correlator. Digital systems [2,4] devote extensive effort to preprocessing. They evaluate the quality of an input test FP and will often reject it (P_R) without any other tests run (based on its quality). In general, digital minutiae matching systems require 15-25 minutiae before recognition is attempted (with high confidence), else the test FP is rejected at this second level of quality analysis. P_R can be 5% if a high P_C is desired on those FPs evaluation [M. McCabe, Pers. Corresp. 5/02]. It typically takes 2s to process an FP and extract is minutae; matching this to 175K people (1.75 million FPs) takes about 40s (approximately 23 µs per FP).Optical correlator systems do not, in general, perform extensive FP image preprocessing. Some papers claim better results with binary FP inputs [24-25], while others note better results with gray scale FP inputs [27]. In our work, we binarize the test input (to allow fast morphological processing steps).

These FP issues are also of concern in digital processing, but again no extensive organized tests have been performed to address such issues.

Most optical FP recognition work has used the joint transform correlator architecture. We consider a frequency plane correlator architecture, as have others [5,6]. The reason for this optical correlator architectural choice is that it is very compatible with use of an optical holographic memory containing all reference FPs [1]. Our results are of use in any optical correlator architecture (with modifications). We also consider filters with complex values, these can be implemented optically by methods noted elsewhere [13,14] or they can be implemented digitally.

Much other optical correlator FP research considers use of phase only filters or devices with coupled phase and amplitude levels for which other encoding methods [5,27] exist. Our results are thus more general and are of use in optical and digital realizations.

2. TEST PROCEDURE AND ELASTIC DISTORTION EFFECTS

The following test and evaluation procedure has been used by NIST for all FP tests in all prior work and for tests on other biometrics. We consider *verification* (a person enters his FP and a PIN, personal identification number, etc. indicator of who he is) and *identification* (a person enters only his FP) applications. In *verification*, the input FP must match the filter for the stated person with a high filter output above some threshold T. Thus, the output from only one filter would be analyzed. In *identification*, we address whether the person is in the database (does the input FP give a high output above T for some filter in the database). In this case, the output from all filters would be analyzed. In *evaluating an FP recognition system* for both applications, tests are different; we also test all other filters in the database (one per person) vs. each input. For verification, any other incorrect filter outputs above T are false alarms in our evaluation procedure. In identification, the correct filter output must be the largest (if not, this is a false alarm) in our evaluation procedure. This stringent evaluation procedure ensures that such tests have a high confidence. For each test FP in identification, only the filter with the largest output (above T) is considered in scoring P_C or P_{FA}. Thus, in evaluating both verification and identification performance, we do not have to test the system vs. a number of input prints not in the database (such inputs should not give outputs above T).

Input tests FPs can/will also be rejected (P_R is the percentage of the test FPs rejected, no decision is made on them). In standard minutiae matching FP recognition, many rejects occur, since these systems only work well on good quality FPs. If the test FP is determined to be of poor quality, it is marked for manual analysis. A major advantage of distortion-invariant filters for FP recognition is their ability to operate on poorer quality test FPs and to not require any FP image enhancement. P_R results are thus included in our verification and identification evaluation scores as we now discuss. In identification of an input test print, if no filter gives an output $\geq$ T, then that test FP is rejected and counts toward P_R. Thus, for identification the number of prints correctly recognized, plus the number falsely classified plus the number rejected equals the test set size. Thus, a test print that gives *no* filter output $\geq$ T is a reject. For P_R in verification, we look only at the correct filter output for a test print, if it is not $\geq$ T, that test print is rejected. The reject procedure is different for verification and for identification, since in use all filter outputs are examined in identification and only one filter output is examined in verification. All P_R scores are a percentage of the test set.

We now detail the *evaluation procedure* we use for a given system for FP verification for a database of 50 people, assuming no tests on additional false FP inputs (not present in the database). We assume a database of 50 people (FPs) with 50 test inputs (test set), one per person, and a set of 50 filters, one per person. We correlate each test input vs. all 50 filters and enter the correlation peak values in Table 1 (2500 entries). The true correlations (correct persons) lie along the diagonal. If a correct filter output is above some threshold T, that input FP is recognized; if any other filter output is above T, it is a false alarm. For this 2-D array, we reduce T and consider ALL entries in the table above T. Entries along the diagonal are correct classifications, ALL other entries above T are false alarms. *Note that a given test FP can give an output above T for several filters and can thus contribute several false alarms*; these false alarms can arise even if the correct filter gives the largest output. We continue to reduce T and repeat the scoring until the correct filters for all 50 test inputs exceed T; this will give us data up to P_C of 100%. The same T is used for all filters. Thus, as we reduce T, false alarms can occur for a test input that was correct (the proper filter output was above T and was the largest). *A single poor quality test input can easily require a very low T and result in many false alarms* (more than just 50), but *the threshold T will be chosen so that no decision is made on such poor test inputs*, i.e. it is rejected and the user is requested to reenter his FP. As T is reduced, we obtain an ROC (receiver operating characteristic) curve of P_C vs. P_{FA}, from which we select a threshold T operating point (usually for a desired small P_{FA}). In this case, at a given T (for a database of N FPs), we say that for verification *the system can recognize P_C% (true positives) for all N FPs with P_{FA}% errors (out of N^2-N possible errors, these are false positives), while rejecting some P_R% of the test prints (false negatives) for which the true filter output is not above T* (new test FPs are needed and requested for these rejected FPs, as they are typically not of sufficient quality). For a given T, entries along the diagonal $\geq$ T contribute to P_C, entries along the diagonal < T contribute to P_R and off-diagonal entries $\geq$ T contribute to P_{FA}. For our example of a 50 FP database, the maximum number of false alarms possible is 50^2-50=2450; thus P_{FA} is a percentage of 2450.

After training has been completed and the threshold T has been chosen, on-line verification tests would only consider one filter and if its output is above T (the output for the FP class stated by the user). We will not normalize each column in Table 1 with respect to the autocorrelation of the input test print for the corresponding filter (for a given filter or person to be verified). This was considered [10] to allow P_C = 100%, but it yields too many false alarms when an input print is poor and normalizing a column assumes that the correct filter is known [10]. In our identification and verification tests, the same threshold T is used for all test inputs and for all filters. The reasons for this is that the threshold cannot be predicted in advance. This is true for these initial tests (using only training set data). However, after several sets of tests, one obtains an idea of the T level to expect for both true and false inputs and thus one can also obtain an indication of this expected level for T (it could then be adjusted for problematic FP filters, determined by the number of training images needed, etc.; these initial tests did not address this advanced issue). Thus, all initial tests used the same T level for all filters.

The *evaluation procedure* for identification applications is now addressed. In this case, on-line tests for each input FP consider all filter outputs; any filter with an output above T is acceptable. Thus, evaluation tests of such filters must ensure that the correct filter gives the largest output with a high confidence; this is necessary to provide high confidence for the identification test results. *The identification test evaluation is more stringent than the procedure for verification* (since P_C = 100% is possible only if P_{FA} = 0% as we detail below). Use of more stringent evaluation tests is logical since the user provides less data (only his FP and not his identity). In the on-line use of an FP identification system for access control, if any filter gives an output $\geq$T, access is granted. Thus, in the evaluation of such a system, we wish to ensure that the correct filter has the largest output with a high confidence. Hence, our test procedure considers all filter outputs and a different P_{FA} measure as detailed below.

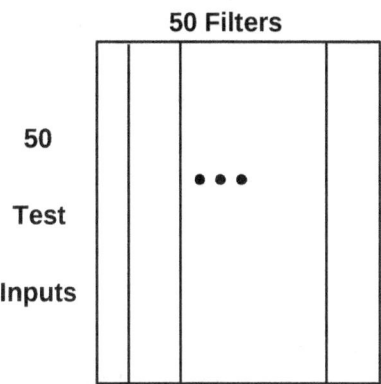

Table 1: Verification test procedure for the case of 50 fingerprints (and no added false class test fingerprints not in the database).

The test procedure we use to evaluate a system for FP identification tests for 50 FPs or filters (one per print) is now discussed. The inputs are the 50 print test set (one per FP). Table 2 shows the output data for this case. We correlate the 50 test FPs with each of the 50 filters, and record the correlation peaks as in Table 2. Our purpose is to determine if the input print is in the database. We also wish that the confidence is high that the correct filter gives the largest output. Thus, test scoring is now different. We set a given threshold T, but now *keep only the maximum entry in each row* if it is above T. If it is correct, we increment P_C; if not we increment P_{FA}. If no entry anywhere in a row is above T, we enter no score (that FP is a reject and counts in the P_R rate). Thus, $P_C + P_{FA} + P_R = 100\%$ for identification verification tests. To generate an ROC identification curve, we continue to lower T until each of the 50 test FPs has some filter output greater than or equal to T. This allows a maximum of 50 false alarms (one per row). P_{FA} is thus a percentage of 50. Note that perfect identification can only occur if there are no false alarms. P_C is the percentage of all FPs in the database (50) for which the correct filter output is the largest. For identification, P_R is the percent of the 50 test prints with no filter output $\geq$ T. For identification, $P_R = 1 - P_C - P_{FA}$. P_{FA} is the percent (out of 50 possible cases) in which the wrong filter gave the largest output for a given test input. For identification, for a given T and N test FPs, we say that *the system recognizes P_C% (true positives) of all N FPs with P_{FA}% errors (only the largest filter output $\geq$ T is considered for each test input) with P_R% rejections (no filter output is $\geq$ T)*.

Fingerprint identification tests are more stringent than for verification tests *in evaluation*, since the on-line tests for verification (any output above T will admit you) are less stringent.

We will not separately normalize each row of Table 2 (as was initially considered) [10] with respect to the autocorrelation peak value for each true and false test input; this was considered to allow improved detection of poor test input FPs.

Figures 1 and 2 vividly quantify the significance of elastic distortions in terms of the differences they produce for the same FP. Figure 1 shows two versions of the same print for different dabs on a real-time input FP reader with different elastic distortions present. One would say that both prints look very similar, but this is not the case. To demonstrate this, we show in Figure 2 the location of 10 minutia (their locations in the FP in Figure 1a are indicated by a black +, their locations in the FP in Figure 1b are indicated by a white +). As seen, different parts of the print are

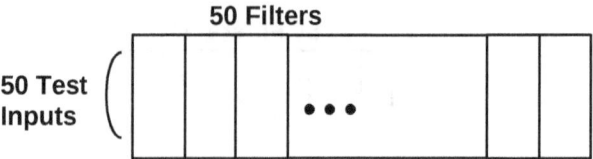

Table 2: Identification test procedure for the case of 50 fingerprints.

shifted by different amounts and in different directions (the top of the print is shifted right, the regions on the right side are shifted slightly left, the bottom part is shifted noticeably to the left, etc.). These shifts (elastic distortions) are quite significant. Many are more than 1-2 ridge widths and thus greatly degrade a correlation output. This vividly shows the effect of elastic distortions. They represent severe distortions. The mean square error overlay of the two images in Figure 1 shows that less than half of the image pixels are similar. In most elastic distortion cases, the cross-correlation of the original and elastic distorted FP show correlation peak losses over 90%. Overcoming elastic distortions is thus a major new result and aspect of our work. Elastic distortions arise when the person dabs his FP onto some FP reader. Differences in pressure between dabs and nonuniform pressure result in very large elastic distortion differences. These differences are much larger than those that arise in rolled FPs, most databases are of rolled FPs.

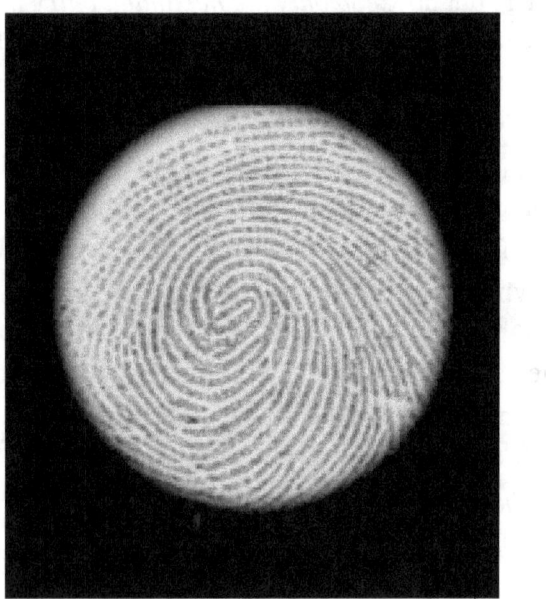
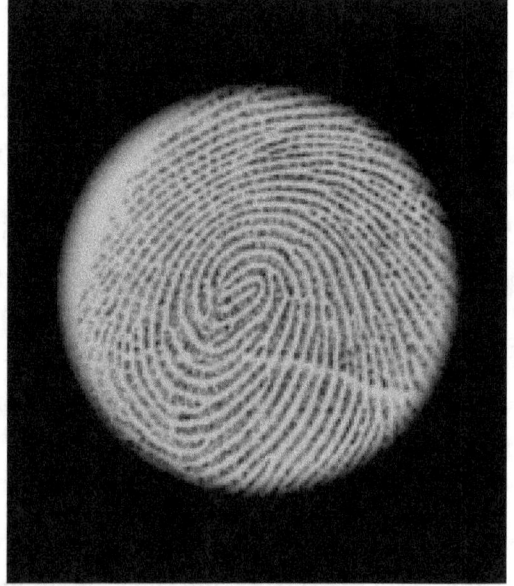

Figure 1. Same print with different elastic distortions.

The issues to address are:

(1) To quantify the effect of elastic distortions on a correlator.

(2) To detail the instructions needed of a user during training when FP samples are requested for filter synthesis.

(3) To address if elastic distortions can be handled in a correlator using distortion-

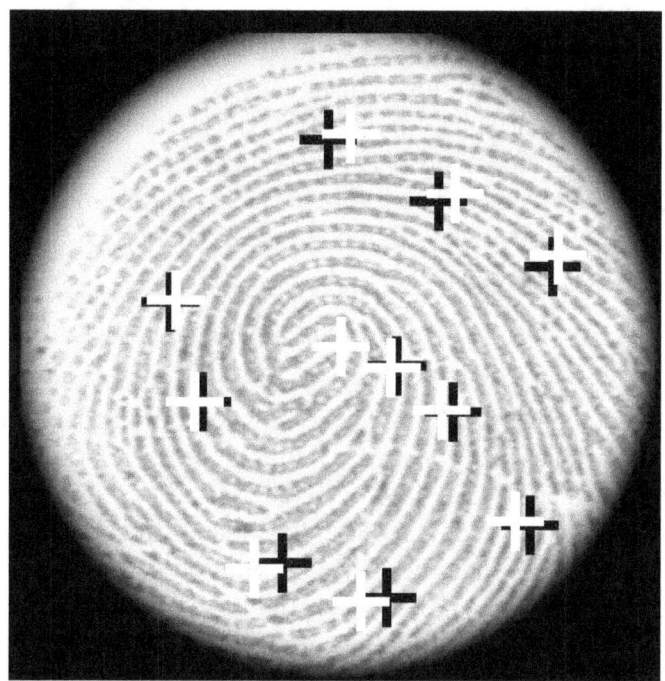

Figure 2. Locations of minutiae shifts in Fig. 1 due to elastic distortions.

invariant filters.

(4) To compare the performance of commercial minutia matching systems to that of a correlator.

Later issues include:

(5) To quantify holographic memory and SLM (spatial light modulator) noise and degradation effects.

(6) Details of the holographic memory architecture: are only several detectors OK, is no input SLM OK, shift invariance, etc.

This initial effort has shown that elastic distortions are very severe and that they can result in the loss of 90% of the autocorrelation peak value. Distortion-invariant filters seem to be able to be fabricated to address such distortions. A better FP database should be acquired to produce more useful samples per print with less FP area omitted. A larger FP database for test results is also needed.

3. GENERAL FILTER SYNTHESIS REMARKS

The distortion-invariant filters we consider are weighted combinations of training set images. For our immediate application, the training set images represent versions of each FP with different elastic distortions present. The distortion-invariant filter is designed to recognize the training set of images and it is thus expected to generalize to recognize test set images (if they are representative of the training set data). Many distortion-invariant filter algorithms exist [11]. Emphasis is generally on optical realization. Thus, many are intended for realization using filters or inputs with only binary or ternary phase levels. We consider new versions of Minace (minimum average noise and correlation plane energy) [12] and other filters with multiple amplitude and phase complex frequency plane values. These filters are thus of use in digital as well as advanced optical realizations (that allows other encoding methods to convert complex data to real data for encoding on a real-valued optical device [13-14]).

The Minace distortion-invariant filter [6] algorithm designs the filter as a vector $\underline{H}$ in the frequency domain. It satisfies peak constraints $X^H \underline{H} = \underline{u}$, where $\underline{u}$ is the unit vector and the rows of the transposal conjugate data matrix $\underline{X}^H$ are the conjugate Fourier transform of the training set images included in the filter; this *peak constraint* requires the correlation peak to be one for all images included in the filter. The Minace filter is also required to minimize the correlation plane energy due to each training set image; this reduces low spatial frequency FP data, this improves rejection of other false FPs and produces narrower correlation peaks (since the spectrum of the FP is whitened by this operation). To provide better recognition of distorted versions of the true FP and of noisy versions of the true FP, the filter also minimizes the correlation plane energy due to white Gaussian noise (characterized by a diagonal matrix $\underline{S}_n(0,0)$). We combine both of these objective functions to be minimized by forming the matrix $\underline{T}(u,v) = \max[\underline{S}_1(u,v),...,\underline{S}_N(u,v),c\underline{S}_n(0,0)]$ at each spatial frequency (u,v), where $\underline{S}_1$ to $\underline{S}_N$ are diagonal matrices whose diagonal elements are the magnitude Fourier transform squared of each training set image included in the filter (they describe the minimization of correlation plane energy). Minimization of the last term $\underline{S}_n$ minimizes correlation plane energy due to input image distortions. We normalize the dc value of each training image to one. The solution for the filter $\underline{H}$ that satisfies the peak constraints and minimizes the combined objective function has been shown [12] to be $\underline{H} = \underline{T}^{-1}X(X^H\underline{T}^{-1}X)^{-1}\underline{u}$. The parameter c included in $\underline{T}$ is a free parameter chosen for a given application (its choice is discussed in Section 7.2 for the present FP database). Large c values emphasize recognition and low c values emphasize rejection of false alarms. An approximation to this c preprocessing that we use for other filters is addressed in Section 6.

Using other results [12], it is easy to show that minimizing the objective functions and hence the choice of c and $\underline{T}$ provides a spatial filter preprocessing of the input test data and of the training set data included in the filter (this preprocessing attenuates low frequencies and emphasizes high frequencies, with the c choice determining the emphasis). This preprocessing is vital to reject false FP inputs (all FPs have similar dc and low spatial frequency values, and thus these frequencies must be reduced to improve rejection of false FP inputs, as we will quantify). We approximate this spatial filtering in Section 6 and apply it to other simpler filters in comparing the performance of various filter types for recognition of elastic FP distortions.

In filter synthesis, the training set images included in the Minace filter are typically selected as follows. The filter is first formed from one training image. This filter is correlated with the training set, the training set image with the lowest correlation peak value is added to the filter (a new filter is formed). This procedure continues until all training set images give correlation peak values above some T_{SYN} level (in filter synthesis). It is not desirable to include all training set images in the filter [16,17]; if this occurs, it says that all training set images are significantly different; in such cases, we do not expect such a filter to generalize well to recognize test set images. If this occurs, we increase c, since this reduces high spatial frequency emphasis and will thus improve recognition of training set data and reduce the number of training set images included in a filter. Use of a validation set to select c has been found useful [16] and was employed in this FP application. Our FP filters are not zero-mean; such filters are possible and have been shown to improve P_{FA} in other applications [16,17].

We consider Minace, average, and synthetic discriminant function (SDF) distortion-invariant filters (see Section 7.2). In our initial versions of these filters, all training set images were included in all filters, since the number of possible training set images per class (FP) was small. Future work will address variations of this initial filter synthesis method.

4. NIST DATABASE 24

There are many FP databases (www.itl.nist.gov/div894/894.03), but they are generally all of rolled prints and only contain (at most) two variations of each print. When FPs are input to a modern system for verification, or identification, they will use modern FP readers such as total internal reflection live-scan devices. The user will dab his FP onto this scanner once or several times and these FPs will be processed. These dab inputs for a given FP represent various elastic distortions (with the variations being determined by the amount and distribution of pressure applied across the FP). There is no known extensive database of elastic distorted prints. There are no published test results of the performance of various algorithms on such elastic FP distorted data. *The range of distortions possible with live scan FP inputs (vs. rolled prints) is much larger*, since more variations in pressure are possible.

NIST has recently assembled a database of FPs with elastic distortions (NIST Special Database 24). This database consists of 10 sec of data (300 images) for each of 100 prints (10 FPs per person). During the 10 sec, each person was told to roll his finger (with the FP core generally centered); the intent was to produce a wide variety of elastic distorted versions of each FP with different pressures applied. The database also contains a second 10 sec of dab data for each of the same 100 prints. In these images, the person dabbed his print several times at various rotations and with different elastic distortions; the intent was to produce different rotated versions of each FP. From 4-14 elastic distorted dab images per print were present in this last 10 sec of data. Another set of 100 similarly acquired prints exists (with 10 sec of data and with 4-14 dab images per print). Thus, we assume 4-14 dab images per FP are available. The first 100 prints in the first 2 sets are from the same 5 fingers of 10 males and 10 females. The second 100 prints are from the other 5 fingers of the same 10 males and 10 females. *This is the only known database containing many elastic distorted live scan versions of each of a number of different FPs.* Each FP image is 420x480 pixels and was obtained with a TIR FP reader (Identicator DFR-90).

In the first 10 sec of data (300 prints at 30 frames/sec), there is little difference between the images of one FP over 1 sec. In addition, many of these images contain smearing in addition to elastic distortions. The second 10 sec of data (4-14 different dab images of each FP) are more typical of elastic FP distortions. Thus, this second set of dab images of 200 FPs was used as the training set (with one FP image per print withheld as a reference test set image). There are thus 3-13 training set images per FP available to form each filter, with one image per FP used as the reference test set input. For each set of 4-14 dab images per FP, the FP image with the least rotation was centered and used as the test set input; all other images of this FP were digitally rotated and centered (Section 5) to best align them with the reference test set image and these were our training set. Our training set images thus generally contain only elastic distortions. We expect that, for many prints, the training set is too small to typify the range of elastic distortions possible per FP. Our tests (Section 7) address *how many dab images per FP are to be requested to adequately describe elastic distortions for a given FP.*

For all FP images, the FP image did not occupy the full 420 ×480 pixel area (Figure 3a). For most FPs, the FP is not uniform, is not centered (a portion of the FP is not present even in the full 420×480 captured image), and the FP background is white (Figures 3a and 4a). Nonuniform FP images are expected due to pressure, grease, etc. effects. The other variations consistently occurred and thus represent issues that must be addressed. To reduce the effects of these problems, we set the FP background to zero (black) (it does not then contribute to the correlation peak value), we centered and rotated each print to a nominal view (Figure 4b), and we extracted a 350 pixel diameter circular region of each print (Figure 4c). The resultant 350×350 pixel image in Figure 3b is one of the better ones. The result in Figure 4c is more typical. As seen, a noticeable portion of the print is missing in Figures 4c vs. Figure 3b. For the 200 FP database, the final 350 pixel diameter images contained *noticeable missing FP parts with 70% of the images having over 30% of the FP missing* after centering. Some individuals consistently had most of their FP images as in Figure 4c. Only very few individuals had even 50% of their images of a FP without the missing FP data problem noted in Figure 4c. This is a significant but practical issue of concern.

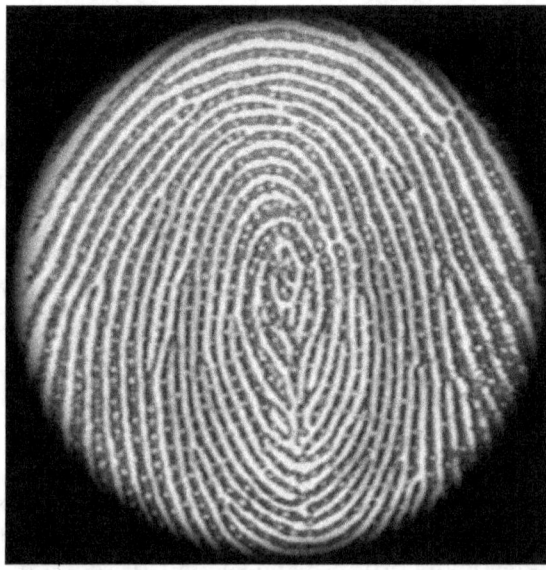

Figure 3a. Fingerprint before segmentation   **Figure 3b. Fingerprint after segmentation**

Our present tests address filters that can perform recognition with such FP database information losses. To reduce this effect in our present database and tests, *we normalize each 350 pixel diameter training (and test) set FP image* by the square root of its energy (thus, even partial FP images have the same energy as full FPs). This normalization also aids recognition of dry and oily FPs as we discuss later in Section 7.3.2. To reduce the spatial frequency effect of the sharp circular aperture, the gray levels in the outer 25 pixels of the aperture were linearly tapered in radius from gray-scale level 255 to 0 in all data samples. Thus, *our initial FP recognition tests are quite impressive as they contain elastic FP distortions (Figure 2), FPs with weak regions and with missing regions near the aperture boundary (Figure 4c)*. Improved dab FP database collection methods are possible (a new database should be produced); but our present tests indicate that advanced filters can handle such practical elastic and other livescan FP distortions. The details of our database preprocessing are provided in Section 5.

5. DATABASE PREPROCESSING PROCEDURES

The FPs have shift, rotation and elastic distortions (in general, all three are present for each FP). They also have bright background regions that must be eliminated, as they contribute to the correlation output. We now detail the preprocessing performed. As noted, our initial goal is to produce a database of FPs for distortion-invariant filter synthesis that have elastic distortions (with other distortion effects reduced).

5.1 Set Background to Zero

For all images (training and test), the white background regions with no FP data (Figures 3a and 4a) were set to zero (black) prior to filter synthesis and before other preprocessing. This was needed to avoid high (white) values in the FP image background that will contribute to the correlation peak output in filter synthesis and in filter tests. This is necessary prior to selecting which training set FP to include in a filter and prior to evaluating whether a test input FP is in a given database. To set the background to black, we first used morphological processing [18] to fill in all of the white regions between the dark ridges in the FP region; this makes the FP region all black. We now detail the preprocessing for the case of the input FP in Figure 5a. We first note that when Figure 5a is shifted and rotated, the image in Figure 5b results, i.e. the background beyond the extent of the original image appears black. Thus, we apply the morphological processing to the original image before shift and rotation, as we now detail.

In Figure 6, we show all steps in this background preprocessing. We threshold the input FP image (Figure .A) at a gray scale level (GSL) of $T = 180$. This GSL was found from tests on several images; its choice is not critical; the same T was applied to all training and test set FPs. The thresholded image is shown in Figure 6b. Binary morphological operations on a binary image are very fast digitally (or optically) and thus thresholding and use of binary morphology are practical and efficient steps. We applied a set of 10 dilations (each with a 3×3 structuring element) to the binary image in Figure 6b. This fills in white gaps between ridges that are up to 10 pixels wide. This was sufficient for all 200 FPs tested. The FP is now black, but its boundary and the boundary of the image have grown by 10 pixels (Figure 6c). Thus, we applied a sequence of 10 erosions each with a 3×3 structuring element to the final dilated image in Figure 6c. The result (Figure 6d) keeps the inside of the print black, restores the proper output boundary of the FP and removes the black boundary grown on the outer edges of the image. A single morphological dilation with a 20×20 structuring element, followed by a single erosion with the same structuring

Figure 4. Typical input original print (a), after background, rotation and centering correction (b), and after circular aperturing (c).

element (this is a morphological closing with a 20 ×20 structuring element) could have been performed and would yield the same result. Such larger structuring element morphological operations are possible optically with no effect on speed [19], however these morphological operations were performed digitally (in this case, performing ten separate morphological operations is faster than performing one large kernel one, using present DSP (digital signal processing) hardware. The resultant binary image in Figure 6d has the FP region black and the

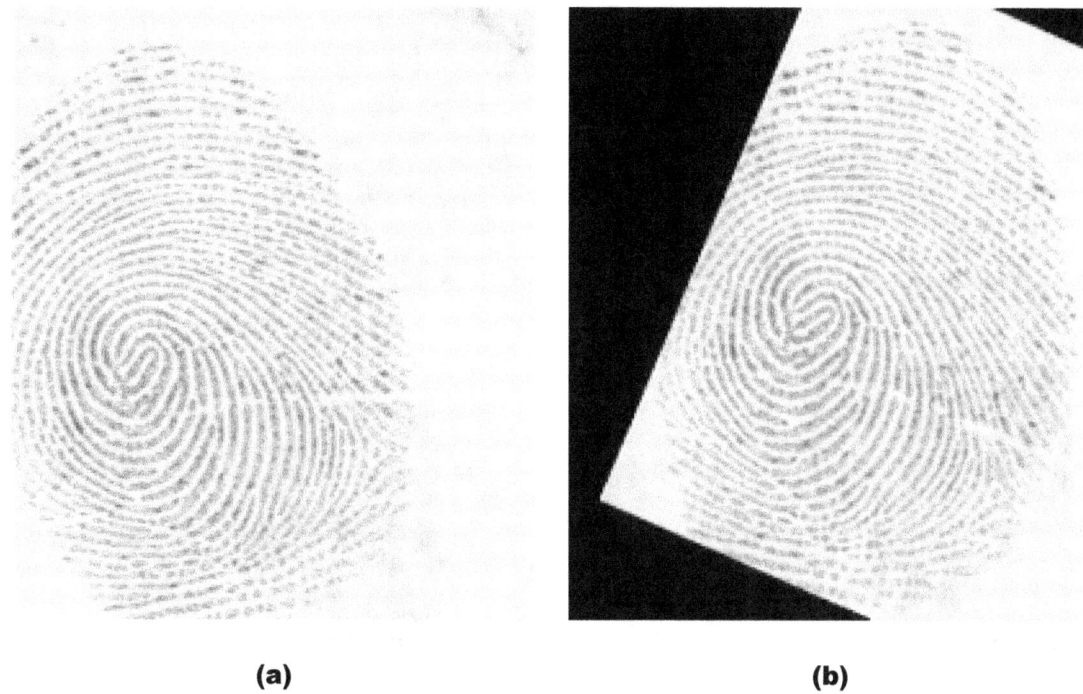

(a) **(b)**

Figure 5. Original FP (a) and the image of it (b) after shift and rotation (note that background beyond the extent of the original image is black after a shift or rotation).

background region white. We contrast reversed this image (Figure 6e) and used Figure 6e as a mask applied to the original gray-scale FP image of Figure .A. This produces an FP image (Figure 6f) with gray scale FP data on a black background. The FP image in Figure 6f is then shifted (centered) and rotated (to a reference orientation) as in Figure 6g. Note that the background is all black, including the portion in Figure 6g to the left of the FP (not present in Figure 6f). The gray levels in the outer ≈20 pixels of the FP were tapered (gradually reduced to zero) to reduce the effect of the sharp edge due to FP data not present in the original image (on the left in Figure 6f). The central 350 pixel diameter region of Figure 6g is extracted, the gray levels in the outer 25 pixels in the 350 circular aperture are tapered and the image is then normalized (in some tests) to produce the final 350×350 pixel image in Figure 6h. The tapering of the background edge data (Figure 6g) and the FP edge data (Figure 6h) was achieved by averaging (blurring) the boundary data and then linearly tapering the pixels in this boundary region from one to zero. Note that the effects of dirt on the scanner (upper corners especially) is not present in the final apertured image.

For completeness, we note that the training set images were high pass filtered for our SDF and average filters to improve discrimination (Section 6). In this case, the 350×350 pixel image with a circular aperture (Figure 6h, but without normalization) was used. Its FT was formed, high pass filtering was performed (Section 6) in the FT domain, the resultant filtered image was transformed back to the spatial domain (350×350 pixels), and then the central 350 pixel diameter region is normalized (in some cases). Note that normalization is performed after high pass filtering. Note that no test inputs are high-pass filtered. They could be, but this has no effect. To provide linear correlations and valid correlation plane values around the center of the correlation plane, the FP images are zero padded in the 350×350 pixel image domain (to 700×700 pixels) *after* high-pass filtering, and normalization (only the central 350 pixel diameter region is normalized). Our filters are formed from these 700×700 pixel zero-padded images. The test set data like Figure 8g are also

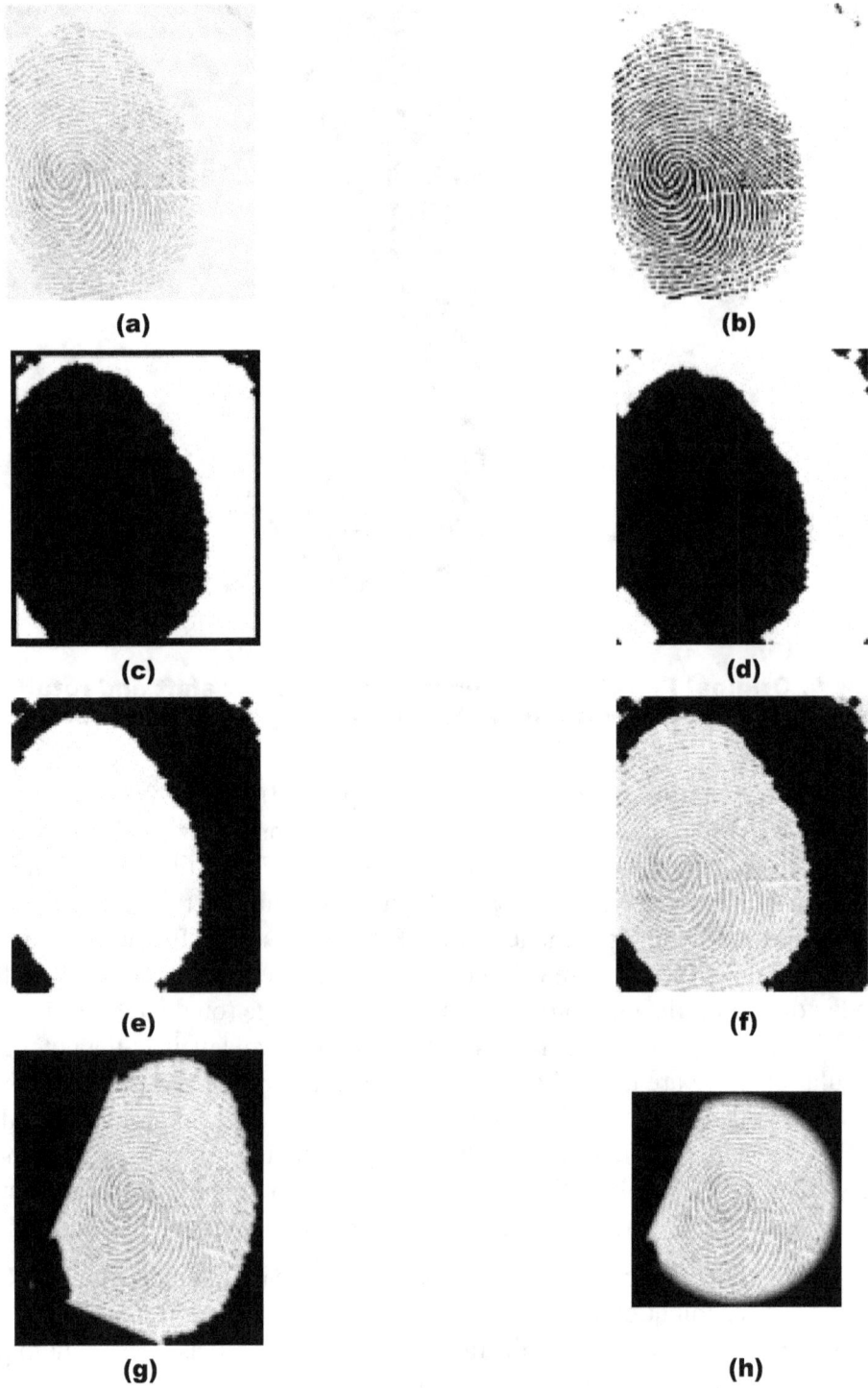

Figure 6. Morphological processing steps to produce a cropped fingerprint with zero (black) background and with aperture edge effect reduced. (a) input image, (b) thresholded (T= 180) binary image, (c) dilation of (b), (d) erosion of (c), (e) reverse binary mask in (d), (f) pass image (a) through mask (e), (g) shift and rotate (f) and taper edges of FP image, (h) crop image in (g) to 350×350 pixels and taper edge of image.

zero padded to 700×700 pixels, the central 350 pixel diameter of the FP in Figure 8f is correlated (after high-pass filtering) with the reference print. Both coarse and fine alignment are used (Section 6) to obtain Figure 8g. For coarse alignment, the full FP in Figure 8f (after high-pass filtering, but with no normalization) is correlated with the reference FP to determine the best shift and the best coarse rotation. To obtain the best fine alignment (shifts and rotations), only the central 350 pixel diameter region of Figure 8f (after high-pass filtering) is used, it is normalized, zero padded and correlated with the reference FP. Each correlation shift is applied to the full image (Figure 8f) and a new 350 pixel diameter region is used after high-pass filtering, normalization and zero padding to obtain each correlation plane value.

We initially wish to assess the effect of only elastic distortions. Thus, we must remove translation and rotation effects. Recall that FP images from the training set are added to the filter if the training image appears very different. We do not want this difference to be due to a simple shift. Various preprocessing was performed to address these new filter synthesis issues, as we now discuss.

5.2 Shift Effects

The initial versions of each FP were not centered. Thus, we initially centered all FPs using a core location algorithm. Filters formed with centered data are expected to be preferable [16,17], since they have more common information, a smaller region of support, require fewer images to be included in the filter, and perform better.

The core detection algorithm, called R92, that locates the core of the FP was coded [20] from a complicated set of six pages of flow charts in Wegstein [21]. The algorithm uses fingerprint ridge flow data to search for the point of peak curvature that typically occurs at the core of the fingerprint. R92 first selects a set of points where the curvature changes in the fingerprint. Each point is scored by analyzing the ridge flow information around that point and the scores are used to determine what point to use for the core. Once the FP core was located, the FPs were centered at their core.

This algorithm can automatically recenter FPs within ±2 pixels for over 90% of the FP data. However, even ±2 pixel accuracy is not sufficient to remove translation effects as we now discuss. *Recentering (or searching over several pixels near the correlation center) is vital in distortion-invariant filter synthesis* and in test data as we now note. Figure 7a shows a 1-D cross-section of the autocorrelation peak for a FP and Figure 7b shows the same cross-section when a region with R = 30 pixels around the center of the Fourier transform (FT) of the FP was suppressed. Section 6 discusses this filtering and why it is needed. It also approximates the preprocessing used in advanced distortion-invariant filters as discussed in Sections 3 and 6. When the low frequency FT region is suppressed, the spectrum of the FP becomes whiter and its autocorrelation becomes sharper as is seen comparing Figures 7a and 7b. With the autocorrelation peak value for each case set to 1.0, we note that the value two pixels from the peak is < 0.2 in Figure 7b. Thus, if vector inner products are used (between a filter and a training set FP) to determine whether to add that given training set image to the filter, many images will be included in the filter (simply because they are shifted). In filter synthesis, this can cause us to include an FP image in the filter, because is "looks" different, while in practice the difference is only a shift. In recognition tests on an unknown input FP, if only a vector inner product is used, an input FP may be rejected because of the peak loss; while in practice the loss was due to a simple shift. Thus, *correlations are essential*

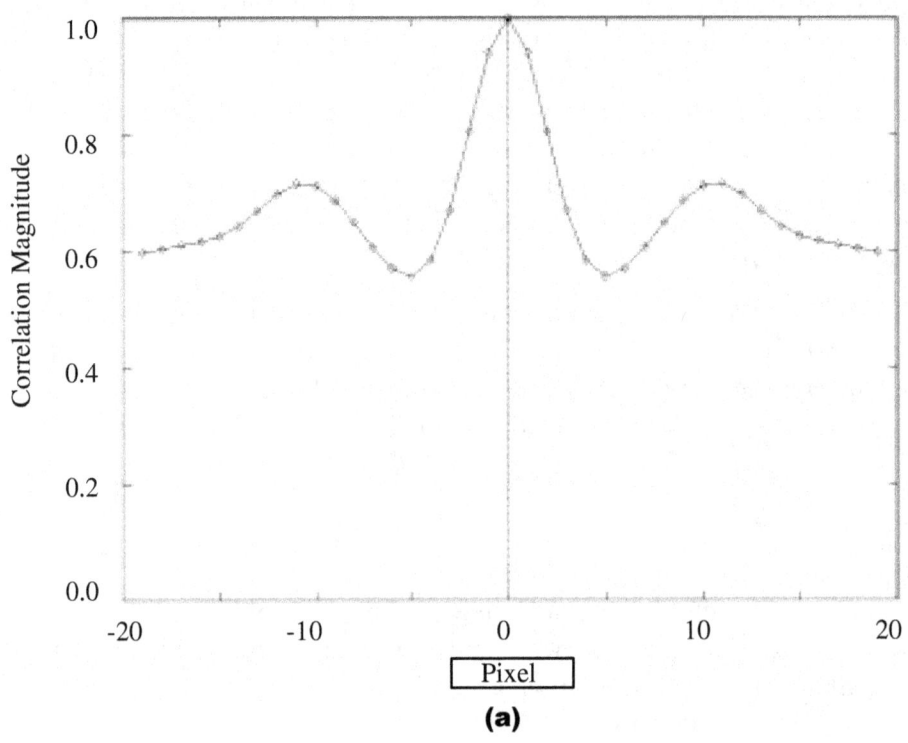

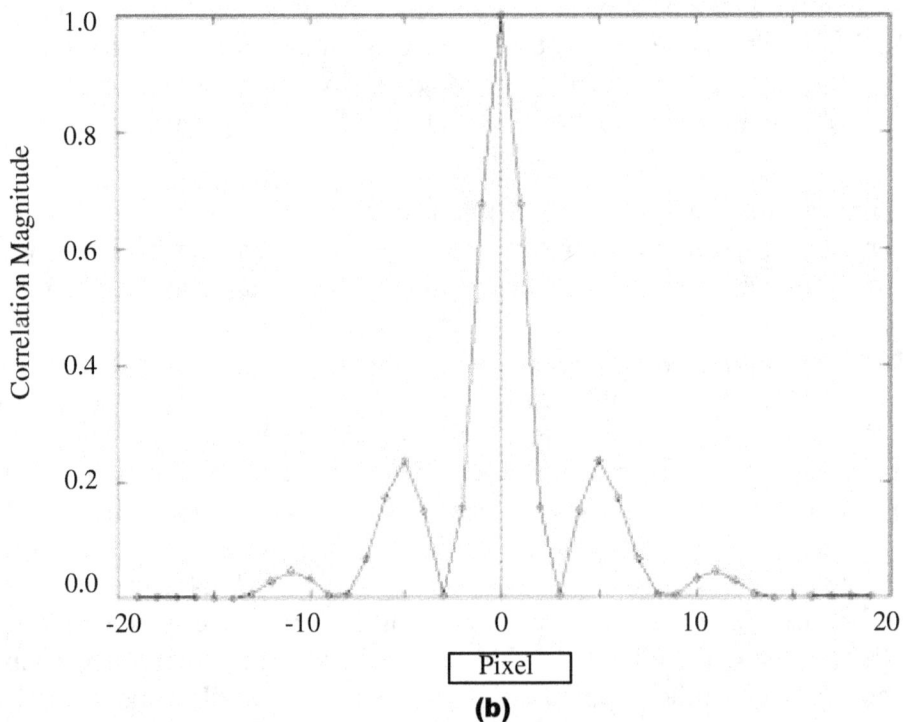

Figure 7: Auto-correlation cross-section with (a) R = 0 and (b) R = 30 high-pass filtering.

and the central region of the correlation plane must be searched both in filter synthesis and in filter tests. We searched a 50 pixel diameter region around the center of each FP. This is larger than is needed, as the largest correlation peak for a filter and a training set FP was always within 12 pixels of the center, when the FPs were initially centered using the R92 algorithm. However, this shows the need to refine the initial FP core detection centering algorithm using correlations. Filters were made and tested, in which the shifts in the FPs included in the filter were based on use of only a small portion of the FPs. These gave much poorer results than when the whole FP was used. Use of small portions of FPs was considered to reduce the correlation time (off-line) required in filter synthesis. However, it is not adequate. With a smaller portion of two FPs used to determine alignment, the elastic distortions greatly affect the alignment shift. With a larger portion (the full 350×350 pixel region) of the FP used, elastic distortions can and do result in a different best correlation shift for alignment. Tests with different area FP images gave shifts that differed by over two pixels. Thus, *training set alignment shifts should be determined from correlations of full 350×350 pixel resolution FP images*. P_C recognition was up to 5% different when filters were formed with training set alignment done using small portions of each FP vs. with the full FP.

Thus, *a correlation with the full FP image is needed to compare prints to determine the shifted version of a training set FP to add to the filter*; a correlation is also needed to locate the largest correlation peak of a filter with an input test print. In filter synthesis, the recommended procedure is to correlate the filter with the training set images, locate the correlation peak, shift the associated training set image so that the correlation peak occurs at the center and include this shifted image in the filter. The procedure we used in these initial FP filter synthesis tests was different. We chose one training set image as a *base image* for shift tests (this was also the test set print). We used the core centering algorithm to initially *coarsely center* all training set images. We then used correlations of the other training set images with respect to the base image for each FP to determine the *fine centering* applied. The fine correlation requires only a small search window and can thus be implemented efficiently digitally by a spatial correlation, rather than by FFTs (fast Fourier transforms). In Section 7, we quantify the improvement provided by fine vs. coarse centering of the training set data and we show that it noticably improves results. During filter synthesis, no further shifts of the training set images were done. *Attention to shift alignment of training set images is a vital new aspect in distortion-invariant filter synthesis.*

5.3 Rotation Effects

All prints now have translation differences removed. The FPs still have rotation and elastic distortion differences present. We use digital preprocessing to remove rotational differences; we expect to apply this to the training set data (to allow more useful training set inputs) and to the on-line test input (to remove/reduce this distortion effect). To provide more FPs to use with only primarily elastic distortions present, the 3-11 training set versions of each FP were digitally rotated (with bilinear interpolation) to the same nominal orientation of the reference test set print. A coarse and then a fine rotation alignment procedure was used. To do this, we located two points on the test (reference) image and the same two points on each training set image (the FP core and a minutia point about 50 pixels from the core were used). We then aligned the FP cores and digitally rotated each training set image to align the two points with those in the reference image, using bilinear interpolation. This procedure was applied separately to the training set for each FP with respect to the associated nominal reference test set FP (base FP image) for each case. The coarse minutia location was manual in these initial results. NIST now has automated software to locate minutia. For fine alignment, shifts and rotations were used (with interpolation) and the largest correlation peak value was used to determine the best alignment.

We expect digital rotations (even with bilinear interpolation) to provide poor accuracy. To quantify the extent of this effect, we digitally rotated one FP by θ degrees, resampled this image with bilinear interpolation, and then rotated it back to the original position, with another bilinear interpolation. Figure 8 shows how the correlation peak magnitude of the original print and the twice rotated FP varies for different choices for θ, with the autocorrelation of the original FP normalized to one. A 20% loss occurs, essentially independent of the amount of rotation. Thus, a single *digital rotation is expected to lose about 10% of the value of the correlation peak magnitude (half of the 20% value)*. This plot was obtained with pixels in a radius of R = 30 pixels of the center of the Fourier transform of all FPs suppressed; this makes the images more sensitive to rotational changes, but is necessary to reduce false alarms as discussed in Section 6. It also produces image data more typical of that expected in advanced distortion-invariant filters. Fine rotation alignment (and shift alignment) is still necessary and is addressed in Section 5.4. We will later quantify the rotation alignment needed for the input for different c and resolution choices.

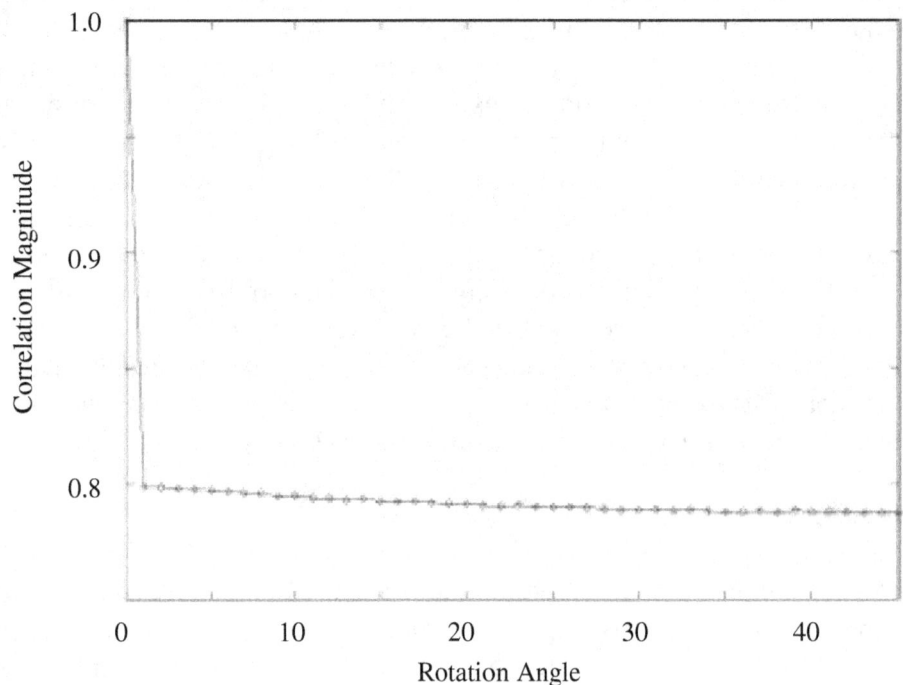

Figure 8. Effects of digital rotation on correlation magnitude.

5.4 Final Corrected/Preprocessed Database

For our final training set images, we digitally corrected the 3-13 training set images per FP for shift and rotation as we now discuss. The R92 core-centering algorithm noted in Section 5.1 was used to provide initial coarse centering and use of the core and a minutia point was used to provide coarse rotation alignment. A finer rotation alignment was made by correlating rotated versions of each training set FP with its reference print. The rotations used were in $1°$ increments over a $±5°$ range. We found additional fine rotations of an average of $2°$ of rotation to be needed and that nearly all FPs required some rotational alignment. The fine rotations are needed, since the location of the minutia used in coarse rotational alignment is expected to be shifted (by definition of elastic distortions) and thus a rotation to align only the core and one minutia point is not expected to be best in terms of rotational-alignment of the full FP. The location of the correlation peak was used to provide the fine shift needed. To produce the final rotated training set images, the FP was fine centered, the fine rotation refinement was added to the original coarse rotation and one combined rotation was applied to the original FP after fine centering; this single rotation was done to reduce the cumulative effect of 10% correlation peak loss in each of two digital rotations. Since the effects of shifts and rotations are coupled, another level of iterations could be considered in future work to jointly reduce both shift and rotation effects. The final 350×350 pixel aperture was applied and normalization was performed after this fine centering and rotation. This leaves only elastic distortions as the dominant distortion. *We still expect a 10% decrease in correlation peak magnitude due to the digitally rotated images* and we must still perform a correlation search to determine the correlation peak for a test FP image.

To coarsely rotationally align each set of FPs, each training set FP is shifted and rotated to coarsely match its reference (test set) FP as follows. The core of both FPs are aligned. The angle θ of the line connecting the minutia point and the core in the training set image is measured and that shifted training set image is rotated until θ is the same as for the reference image. To fine align each set of FPs, this was done with respect to the reference image for each FP (rather than from the filters) using correlation. We formed finely rotated versions of each FP (rotated by $±5°$ in $1°$ increments). We correlated each finely rotated training set FP versus its reference FP, selected the rotated image with the highest correlation peak (this determines the fine rotation needed) and the location of the correlation peak determined the fine shift required. The sum of the coarse and fine shifts and rotations were calculated and one rotation was applied to the original training image (to avoid the composite effect of two rotations on correlation peak loss). The two-step coarse and fine alignment was faster than a single step method since fewer rotated images had to be produced and since the correlation could be performed in space with the correlation evaluated at only the central pixel shifts in the correlation plane. We note that it is essential that the full 350×350 pixel FP image be correlated. When a smaller 100×100 pixel region was used for the correlation (to reduce correlation time), different pixel shifts resulted and all filters performed worse. The fine alignment shifts varied from 0 to 6 pixels with a 3 pixel shift being typical. The fine rotations needed varied from $0°$ to $3°$, with nearly all FPs requiring at least $1°$ of rotation; bilinear interpolation was used with all rotations. Data (Section 7) shows that all filters gave better performance with fine versus coarse alignment. This is expected, since FP ridges are about 4 pixels wide and FP valleys are about 5 pixels wide; thus the sum of two shifted FPs can easily have the ridges of one FP lying on the valleys of another thus reducing the structure present in a filter that is a composite of several FPs.

6. SIMULATED LOW-FREQUENCY FILTERING

Advanced distortion-invariant filters involve preprocessing [12] of the data; this involves suppression of low spatial frequencies (see Section 3). This was included in our earlier digital rotational sensitivity tests in Figure 8 and in our correlation peak width data in Figure 7. For our simple average and synthetic discriminant function (SDF) filters [5], we used FPs with low-frequency suppression, as detailed below. For advanced Minace filters [6], the original images were used (with no low-frequency suppression), since the low-frequency suppression is automatic in Minace filter synthesis.

This section addressed whether low-frequency suppression will improve rejection of *false class* FPs. For false class FPs, we consider prints of the same class (whorl, etc.) as that of the input, since they are assumed to be most similar and to be those FPs expected to provide the largest cross-correlations (false alarms). The thermoplastic filter used in the NIST laboratory demonstration system provides a related low-frequency data suppression; however, we do not attempt to precisely model this laboratory system (since the filters can select which training set prints to include in the filter and since the SDF filter will not equally weight all training set prints). The motivation for suppressing low spatial frequencies is that all FPs have a similar low-frequency spectrum and thus suppressing these frequencies allows one to better address differences between different FPs. The assumption is that *true class* elastic distortions represent smaller differences than those for *false class* FPs. This assumes that suppressing low spatial frequencies will improve the difference between similar FPs (in the same class), while filter synthesis will accommodate elastic distortion variations of one FP. As noted earlier, we normalize all FPs. For our initial average and SDF filters, *normalization of training set images was done* after such high-pass frequency filtering. This high-pass frequency filtering approximates the pre-processing provided by advanced distortion-invariant Minace filters. Thus, it should also improve the performance of average and SDF filters.

We now address the low-frequency preprocessing used. Figure 9 shows the 1-D version of the final low frequency suppression filter chosen to be applied to the FT (Fourier transform) of each FP. It is centered at the dc spatial frequency $u = 0$. It sets a number of low frequency pixels (60) equal to zero and includes a smooth transition region (10 pixels wide) with a Gaussian change in transmittance. This filter has a radius $R = 30$. We tested various versions of this filter (with different R choices). The width W of the transition region was varied for different R choices. For $R \geq 20$, we used $W = 10$; for $R = 10, 8$ and 3, we used $W = 5, 3$ and 1 respectively.

The validation set we used to select R consisted of: one FP (a loop) as the original FP (it was the reference test set image), eight elastic distorted images of this same FP as our true class (these were the digitally rotated dab images from the training set for the same FP), and 5 images of other loop FPs as our false class. The false class FPs were of the same class of FP (a loop), since we expect them to give higher cross-correlations than FPs of other classes. Figure 10 shows the correlation peak values vs. R for the original test set FP with itself (normalized to 1.0) and versus the eight elastic distorted versions of that same FP (true correlations), and the correlation of the original FP vs. the five other loop FPs (false correlations). The largest correlation peak value in the full correlation plane was recorded. For each choice of R, the correlation peak value of the original FP correlated with the filtered original FP is normalized to one.

As we increase R, we expect the cross-correlation of other FPs to decrease; the correlation of

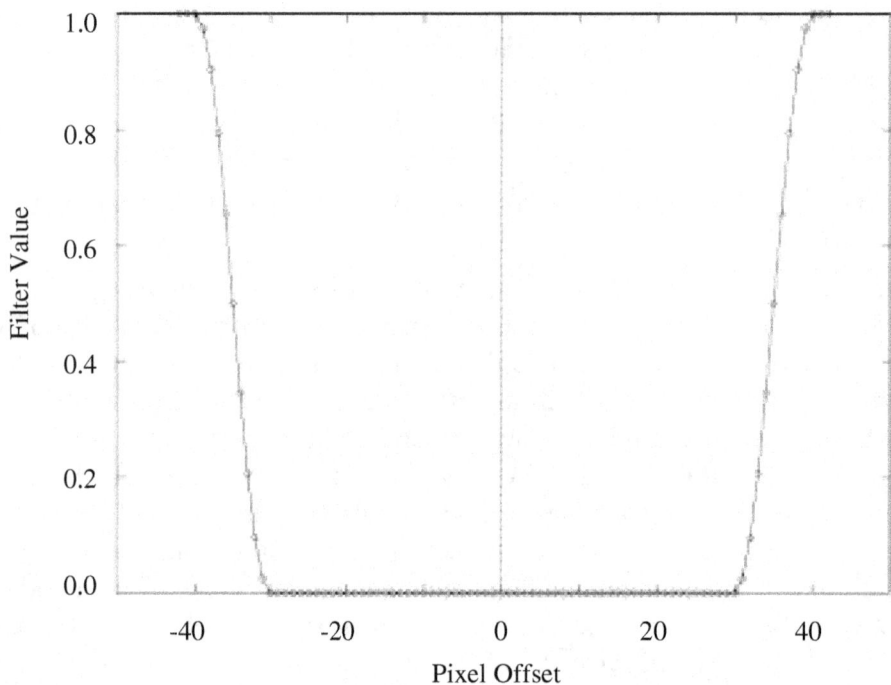

Figure 9. Low-frequency suppression filter used.

versions of the same FP (*true correlations*) will also decrease. We choose R so that the worst (lowest) *true* correlation (between versions of the same FP) is larger than the worst-case *false* correlation (the largest correlation between 2 different FPs of the same type). As seen, all false correlation peaks (0.16 to 0.20) lie below the true correlation peak values (0.25-0.41) for all values of R $\geq$ 5. As we increase R, we see the desired effect that the difference between true and false correlations increases (the difference is fairly constant for R $\geq$ 10 pixels). Thus, *suppression of low spatial frequency data is useful in reducing false alarms*. We chose R = 30 (a wide range of R values are expected to be equally useful) as the high-pass filtering to apply to all of our average and SDF filter data.

In Figure 10a, with the dc value present at R = 0, all correlation peak values are large and decrease rapidly when dc is suppressed; hence, Fig. 10b is included. At R = 0, the true correlation peak values range from 0.8 to 0.91 (the auto-correlation value is normalized to 1.0) and the false correlation peak values range from 0.85 to 0.87. Thus, one cannot separate true and false FPs using the original FP data. This is expected because all FPs have a similar dc average value. *This shows the need for some spatial frequency preprocessing for our FP application.* Removing dc provides more ability to distinguish smaller variations (due to true correlation elastic distortions) and larger variations (due to different prints of the same class, a loop), since these are higher spatial frequency variations. With all FP data filtered with R = 30, if we normalize the original (auto) filtered correlation FP value to 1.0, the true correlations (with elastic distorted FPs) yield peak values from 0.15 to 0.32. This demonstrates the *severe distortions presented by elastic distortions, i.e. correlation peak losses of 70-85%*. The largest false correlation peak value is 0.12 and is below the lowest true correlation peak value for the validation set used in Figure 10. Our

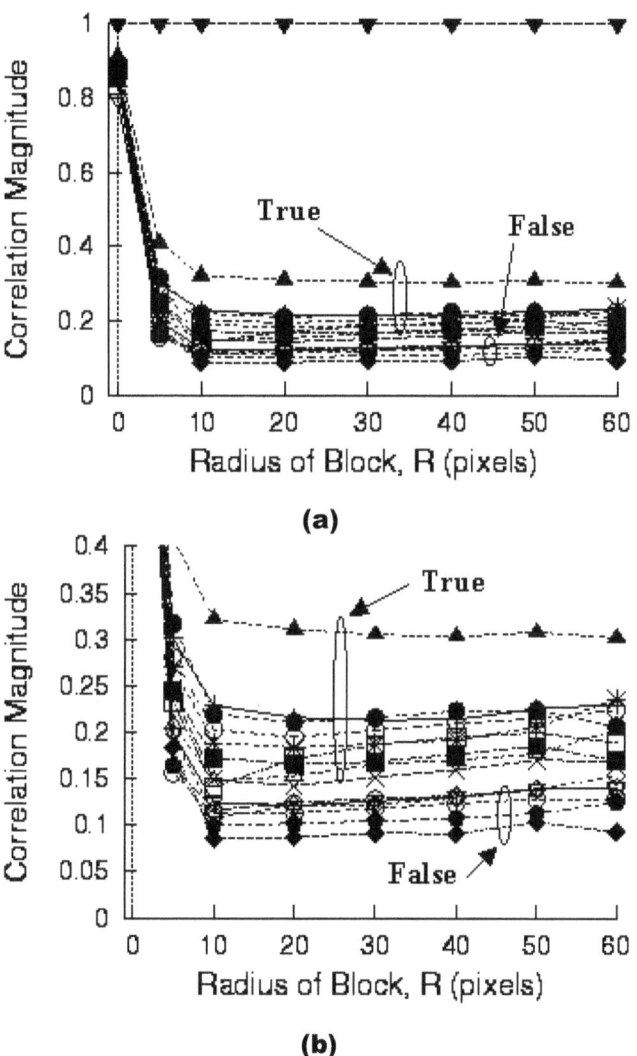

Figure 10: Effect of Low-Frequency suppression on auto, true and false fingerprint correlations. (a) full plots and (b) enlarged scale

filters will bring all elastic distorted correlation peak values (included in the training set) into a much smaller range, closer to 1.0, and provide large true test set peaks (over 0.5 hopefully); thus false correlation peak values should be noiceably lower than true correlation peaks when high-pass filtering is used.

The approximately flat response in Figure 10 from R = 10 to 60 may seem unusual. However, analysis of the FT of the original FP (Figure 3b) shown in Figure 11 explains this result. The FT pattern (Figure 11) is seen to mainly consist of a circular band of data around a spatial frequency corresponding to the ridge spacing of the FP. The low spatial frequency region of the FT is suppressed in Figure 11 to allow the higher spatial-frequency region of the spectrum to be visualized better. This FT pattern is explained by viewing an FP as a dominant spatial frequency carrier, the ridge spacing of the lines in an FP, with the detailed FP information present as modulation around this carrier frequency. The nearly circular band in the FT pattern in Figure 11 corresponds to the spatial frequency of the FP ridges. The variations seen in Figure 11 are the actual detailed FP pattern information on this carrier. As seen, there is no appreciable FP information in spatial frequencies up to a spatial frequency corresponding to $\approx$ 6 to 10 pixels/cycle

(corresponding to ridge and valley widths of 3-5 pixels). At R = 40, the full radius of the low-frequency suppression filter corresponds to a spatial frequency with a period of 17 pixels/cycle; at R = 60, the extent corresponds to a lower period (higher spatial frequency) of 11.6 pixels/cycle. This explains the flat nature of the curves in Figure 10 for R $\geq$ 20. Beyond R = 60, we expect the curves to change. Thus, *selection of R is not critical*.

Figure 11. Fourier spectrum of the fingerprint in Figure 3b with R = 40.

7. TEST RESULTS

No attempt was made to model the optical lab NIST demonstration system (LC (liquid crystal) rise and fall times, TP bandpass filtering, time integration, etc.). The original plans for the NIST database 24 were to use the first 10 sec of data (300 images) as a training set, and the last 10 sec of dab data (4-12 dabs) as a test set. In analyses [10] of the first 10 sec of data, we found (MSE analysis) that over 1 sec there were negligible differences in the FP images. This is expected since human motion is not fast. We also determined that the second 10 sec of data was more typical of the expected inputs (in terms of elastic distortions present), however rotation effects had to be removed digitally to allow use of all available images in these second 10 sec of data. Thus, this second 10 sec of data became our training/test set. We also added the second set of 100 FPs to provide a larger 200 FP database for tests. This was necessary, since we do not expect that the number of training images (3 to 13) is sufficient for most FPs (FPs with only 4-8 images available per FP, would allow for a training set of only 3 to 7 FPs). Thus, the database should be redone to include more dab inputs per print (to better represent the elastic variations expected in an FP).

All results in [10] are not valid for numerous reasons. Each FP was not normalized, low spatial frequencies were not blocked (they were set to 1.0), digital rotation errors were not considered, MSE was used to add new images to the filter, a correlation search of shifts was not done, normalization of each FP was not done, rows or columns were normalized in P_C scoring, etc.

7.1 Database

Our present test of the use of filters to recognize elastic distorted FPs used 10 seconds of FP dab data. Since no person was able to produce more than 14 dab images of one FP in 10 sec, we used the full set of 200 FPs (to provide enough FPs with at least 9 dab images) for our training and tests. There were 55 FPs (out of 200) with at least 9 dab images. These were used as our initial database. We denote the database FPs by A-x, where A = A-G denotes the person used and x = 0-9 denotes the person's finger used. The seven people and the fingers used were ones with at least 9 dab images. We selected one image per FP as our reference (test set) FP image; the original image with the least rotation was chosen as the reference (test) image for each FP. For each FP, the other 8 to 13 images were used as the training set. Initial tests showed that about 8 dab images were sufficient to give a representative set of elastic distortions. All training set FP images were centered, rotationally-corrected using coarse and fine steps, normalized and processed as detailed in Section 5. For all filters, except the Minace filter, the central R=30 pixel region of the Fourier transform of each FP was omitted; this makes the different versions of each FP more different (but increases the differences between different FPs more). Recall that all FPs have elastic distortions present and about 70% of them are missing a large portion (over 30%) of the print (even after aperturing the central circular 350 ×350 pixel region) plus many FPs are oily or dry or scarred. Thus, this is a formidable recognition problem for which some type of composite filter (a combination of training set images) is needed.

7.2 Composite Filter Designs

We tested three types of composite filters. In all cases, we formed one filter for each FP. The simplest was an *average filter*. This filter H (in the Fourier domain) is the sum of the Fourier transforms F_n of the training set images divided by the number N of training set images, $H' = (F_1 + F_2 \cdots F_N)/N$. It uses equal weights for all training set images. This is not necessarily the best combination of training set images, since it assumes that all training set images are equally important. All pixels in H' are multiplied by the same factor to make the largest training set peak equal to one. This yields the final filter H. Thus, the output for partial fingerprints in the training set will be lower and proportional to the fractional area of the useful fingerprint in the input aperture (unless normalized data are used in training, filter synthesis, and testing).

The second filter considered was the *synthetic discriminant function (SDF)* filter. In this case, the combination weights are the elements of the vector a and are chosen to satisfy $a = V^{-1}u$, where V is the vector inner product matrix (its elements are the inner product of each pair of training set images) and u is a vector that is all ones. The SDF filter is thus designed to provide correlation peak values of one for all training set images. The combination weights now depend on the similarity of the different training set images (the elements of V denote this similarity). For the SDFs used here, all training set images for each FP were included in the filter for that FP. This is not necessary and is generally not desirable, but with the small training set size available it is not expected to significantly affect results.

The *average and SDF filters used here are not the standard ones*, since they are formed from high-pass filtered data. Thus, they should have many of the advantages of the Minace filter and should perform better than standard average and SDF filters.

The training set images used in the Minace filter are not high-pass filtered, since the choice of c in the Minace filter selects the type of high-pass filtering performed (lower c values suppress more lower spatial frequencies). To select c, a validation set was used as suggested elsewhere [16]. We used three sets of FPs (each with at least ten dab images). All were whorl patterns. A Minace filter for each was made using 10, 9 and 9 training set images in the three filters respectively. One version of each of the three FP sets was used as test set inputs. An additional six other test set inputs from other whorl type FPs were used to test the c choice for rejection of more FPs. All FPs were of the same type (whorl), since this is expected to present the hardest false class rejection case. For each of the three FP sets, we formed thirteen filters for each FP (each with a different c value from c = 1 (no high-pass filter preprocessing) to $c = 10^6$. For each choice of c for each filter, we calculated the smallest true correlation peak value and the largest false correlation peak value (for each FP filter, there is one true class test FP input and nine false class FP test inputs) in a 50 pixel diameter region around the center of the correlation plane. We do this for all thirteen c values for all three filters. We calculated the difference (*smallest true* correlation peak value minus *largest false* correlation peak value) vs. c for each of the filters and selected the c value that gave the largest difference. Figure 12 shows the difference vs. c for the three filters. As seen in all cases, large differences (above 0.3) occur for all filters over a wide c range (from 10^{-6} to 5×10^{-4}). Thus, the choice of c is not critical. We chose $c = 5 \times 10^{-5}$. We used the same c value for all 55 filters in these initial tests; it is possible to use a different c value for each FP filter.

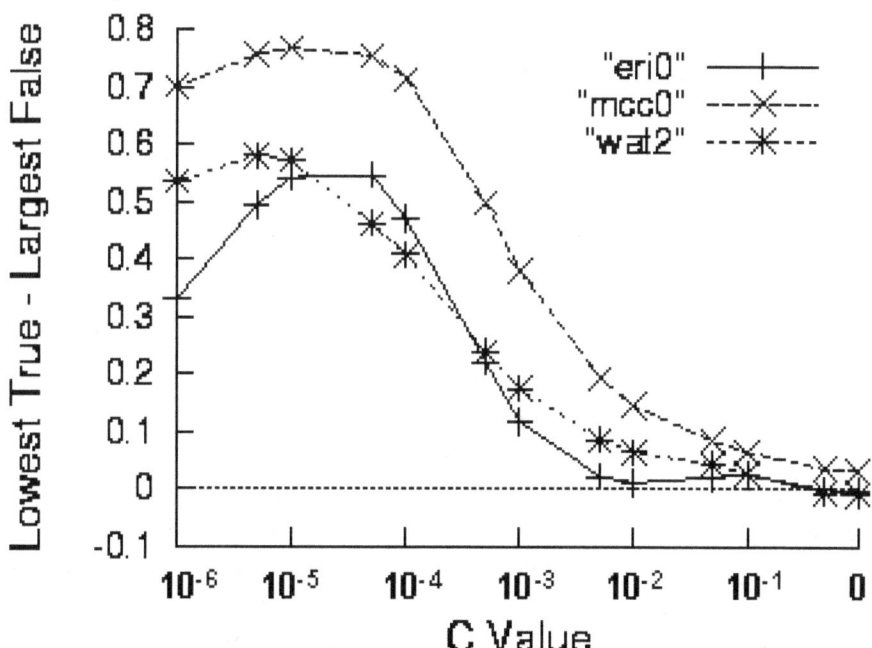

Figure 12. The Minace c parameter choice vs. the minimum true minus the maximum false class correlation values for the validation set of three whorl FPs, each vs. eight other whorl FPs.

In these initial FP Minace filters, all training set images were included in each filter and the fine shifts with respect to the reference image were used (no additional shifts of training images were used in filter synthesis) and a 50 pixel search region around the center of the correlation plane was used to locate the largest correlation peak value for test FP inputs. Analysis of the effect of c on the maximum false peak and the minimum true peak value for two filters are shown in Table 3. As seen, as c is reduced, the trend is for false peaks to decrease (as expected) and for true peaks to decrease only slightly initially and then to decrease more at lower c values (this is expected since at lower c there is less low frequency data and the true test input will differ from the training set images by a larger amount). Thus, use of a low c value (down to some limit before the minimum true peak value starts to noticeably decrease faster than the false peak) improves the minimum true minus the maximum false difference and hence the expected P_C (true class recognition) and P_{FA} (false alarm) results. The maximum false class values (0.26 to 0.27) are much less than the minimum true class values (0.77 to 0.84) at the $c = 5 \times 10^{-5}$ value we chose. Thus, we expect excellent performance for $c = 5 \times 10^{-5}$ Minace filters. Although the correlation peak difference (minimum true minus maximum false) is positive for most cases, the very large false peak values 0.999 at c = 1 (no preprocessing) and 0.9 at $c = 5 \times 10^{-3}$ are so large that they are expected to be larger than many of the other FP test input true correlation peaks (subsequent tests on more FPs show true correlation peak values below 0.6 with our $c = 5 \times 10^{-5}$ choice). *Thus, there is clearly a need for and a distinct advantage in the automated high-pass filtering provided by selection of c in the Minace filter to reduce false FP correlation outputs.* All three of our distortion-invariant filters use some such preprocessed training set data (this filtering is included in the filter and is thus also effectively automatically applied to the test input FP).

Table 3: Correlation peak values vs. c for two validation set filters.

Filter	Correl Value	c = 1	10^{-2}	10^{-3}	10^{-4}	10^{-5}	5×10^{-5}	10^{-6}
Eri0	max false	0.999	0.93	0.79	0.48	0.30	0.27	0.20
	min true	0.994	0.94	0.91	0.95	0.85	0.77	0.53
Wat0	max false	1.01	0.93	0.79	0.49	0.30	0.26	0.19
	min true	1.00	0.99	0.96	0.91	0.87	0.84	0.73

With c selected to be 5×10^{-5}, Minace filters were formed for each of the 55 FPs. In each case, the same c value was used for all FPs, and all training set images were included in all filters (since the number was small, a maximum of 13 and a minimum of 8, we do not expect this to significantly affect results). All training set images were included in all three types of filters; thus, it is fair to compare different filter results. When all training set images are included in a filter, the lowest c value is generally required, since, if we increase c, fewer training set images are generally needed in the filter. Thus, we expect the present $c = 5 \times 10^{-5}$ value to be close to the minimum and that higher c values will be preferable in general and in cases when fewer training set images are present. Thus, our present c choice is only optimum for cases when at least eight training set images per FP are available. For other FPs, with fewer training set images, larger c values are expected to be better.

7.3 Need for Aligned and Normalized Data (Initial Verification Test Results)

All results show P_C (percentage of all test FPs correctly recognized) vs. P_{FA} (number of false test FPs recognized incorrectly) ROC (receiver operating characteristic) data. The P_{FA} axes are shifted negative to allow better display of the perfect performance obtained in several cases. Only the low P_{FA} data portion of the full ROC curve is shown, since this is the operating region of interest. Verification test results are considered to determine if fine or coarse-aligned data are needed and if normalized or unnormalized data are needed. We obviously expect fine-aligned normalized filters to perform better, but we will quantify the differences for the three different filters.

7.3.1 Need for fine-aligned Data

For verification data (Figure 13), recall that the P_C scores are out of a maximum of 55 correct, P_{FA} scores are out of 2970 (= 55^2 - 55) possible errors (false class inputs), and P_R (rejects) are out of the 55 test inputs (only the correct filter output is considered; if it is < T, that test input is rejected). Thus, $P_C + P_R = 100\%$. By comparison, for identification (Figure 15), P_C, P_{FA} and P_R are all out of 55 (due to the maximum scoring mechanism used), and $P_C + P_{FA} + P_R = 100\%$. For verification data, P_{FA} error scores are low, since they are out of 2970 possibilities. A P_{FA} = 0.034% corresponds to one error (at a given T, if any wrong filter gives an output $\geq$ T, that is a P_{FA} error, even if the correct filter output is largest).

We first consider whether use of fine vs. coarse alignment data gives better performance and we quantify the improvement obtained. Figure 13 shows these results for the three filter types for normalized data for the verification evaluation case. Both the training and test set images are all either coarse or fine-aligned in the two cases. Both fine shift and rotation alignment are included in the fine data cases compared to the coarse alignment cases. Fine rotation of the FP data is expected to improve results, since larger true output correlation peaks result when the test data is better aligned with respect to the reference training images, since then the training and test data are more similar. Fine alignment of the training set with respect to itself produces a more common training set and filters with less blur (the combination weights for SDF and Minace filters are different when the training set is fine vs. coarse-aligned); such filters with more structure also result in reduced false correlation peaks. Later, filter reconstructions confirm the better filter structure using fine-aligned data (Section 7.6.6). Figure 13 shows test set results. In all cases, the *ROC data are better when fine alignment of the training set is used*, as expected. The improvement in P_C is generally at least 5% and as much as 25% when fine-aligned training data is used.

7.3.1.1 Minace Data Analysis

For the Minace filter (Figure 13a), the filter with fine alignment has perfect performance (P_C = 100%, P_{FA} = 0% and P_R = 0%). These perfect results occur for a threshold T $\geq$ 0.48 up to 0.56 (the minimum true correlation test peak is 0.56 and the maximum false test peak is 0.47). With coarse alignment, the ROC curve for the Minace filter is worse, P_C = 74.5% at P_{FA} = 0% and P_R =

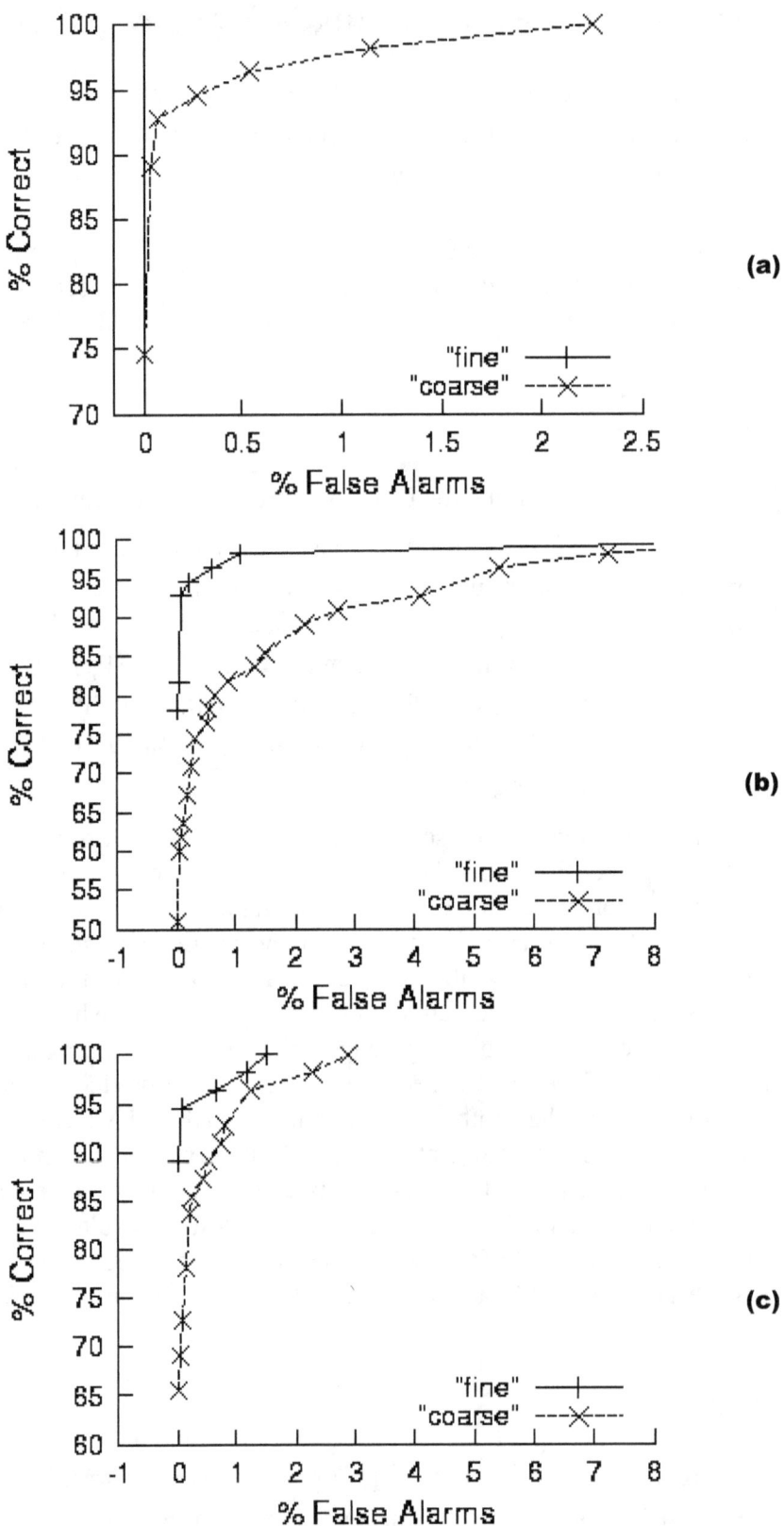

Figure 13. Fine vs. coarse alignment (shift and rotation) improvement using normalized FP data. ROC verification results for the Minace (a), SDF (b) and average (c) filters.

25.5% for T = 0.55. Thus, use of fine-aligned training set data improves the performance of 25% of the filters. For coarse-aligned data, the minimum true test correlation peak is 0.41 and the maximum false test correlation peak is 0.546. For fine-aligned data, the minimum true test peak is larger (0.56) and the maximum false test peak is less (0.48). Thus, use of fine-aligned data increased all test set FP true correlation peaks so that all were ≥ 0.56. This is clearly due to better alignment of the test FP. In addition, use of fine-aligned data also decreased many false correlation peak values (when coarse-aligned data was used, 8 false correlation peaks were ≥ 0.48 and some false peaks are as large as 0.55; with fine-aligned data, no false peaks are ≥ 0.48). *We expect that these improvements occurred because fine rotational alignment of the input is needed and because the associated filter now has more structure with less washout of ridges and valleys.*

With coarse-aligned data, four of the 55 filters do not perform well giving true test set correlation peak values less than 0.5 (two true test peaks are ≤ 0.44); use of these low T values introduced false alarms. Thus, four of 55 filters ($\approx 7.3\%$) need fine-aligned data to provide good test set peaks > 0.5. The largest false class peak (0.546) is very high (one other false peak is also > 0.5); thus, a high T = 0.55 with an associated large number of missed true class test images (P_C is only 74.5%) is needed to obtain no false alarms ($P_{FA} = 0\%$). For coarse-aligned data, we can achieve $P_C = 100\%$ if we reduce T to 0.41 (the minimum true test peak value), but then $P_{FA} = 2.26\%$ (which is very large); at an intermediate T = 0.51 level, $P_C = 89.1\%$ while $P_{FA} = 0.034\%$. For both coarse and fine-aligned data, there were no false FP peaks ≥ 0.55.

For the region of the coarse-aligned ROC curve shown, the threshold T varied from 0.55 at $P_{FA} = 0$ to 0.41 at $P_{FA} = 2.26\%$ and $P_C = 100\%$ (for coarse-aligned data). For verification scores, $P_C + P_R = 100\%$, thus the number of rejected test FPs can be determined.

All errors (with coarse alignment) are clearly due to not properly fine aligning training (and test) set data. As seen, this can produce a 5% to 25% drop in P_C. Thus, *fine alignment of the training set for the database is necessary and vital to obtain best performance.*

It is very important to note that for *all Minace filters* the minimum correlation true test peak output was larger than the maximum false correlation peak (out of all 2970 possible cross-correlations) if fine-aligned and normalized data are used. Similarly, *for a given FP test input*, the correct Minace filter output is always the largest out of all 55 filters; this occurs for all data cases (coarse or fine-aligned and normalized or not). The first remark reflects the case of no errors in verification. The second remark applies to the case of identification. Thus, data alignment is of less concern for identification than for verification (since only the maximum filter output is used in identification tests). For verification cases, the different filters and test inputs could use different T choices to determine if a correlation peak occurred; we do not do this; we use the same T for all cases and thus find some false alarms. This is why fine-aligned data is needed to increase all true test set correlation peaks above some T. *Future work could consider use of a different T for different problematic FPs. In this case, less preprocessing alignment may be acceptable.*

7.3.1.2 SDF Filter Analysis

The corresponding SDF and average filter curves (Figures 13b and 13c) show similar trends for coarse vs. fine-aligned data. For the low P_{FA} values ($\leq 1\%$) of interest, better P_C is obtained (by 9-27%) when fine-aligned data are used. SDF results (Figure 13b) are now discussed. For fine-aligned data, P_C increases by ≈27% vs. coarse-aligned data (from 50.9% to 78.2%) at P_{FA} = 0%, and by ≈21% (from 61.8% to 92.7%) for P_{FA} = 0.067% (two false alarms). Thus, with fine-aligned data, SDF filters give good performance: P_C = 92.7% with P_{FA} = 0.067% (only two false alarms). For fine-aligned data, the minimum true correlation peak is 0.25 (this is very low) and the maximum false correlation peak is 0.55 (this is very large). For coarse-aligned data, the minimum true peak is 0.27 and the maximum false peak is 0.55. These two limits are similar for both cases, however the ROC data in Figure 13b are noticeably better for the fine-aligned data case. Fine-aligned data is thus desired and is now discussed.

For the fine-aligned data case, at T = 0.56, there are no false alarms (the largest false peak is 0.55); thus we cannot expect to operate at P_{FA} = 0 and achieve a very high P_C with this SDF filter (there are ten true class test peaks < 0.5). As T decreases from 0.56 to 0.395, P_C increases from 78.2% to 94.5% (52 of 55 FPs are correct) and P_{FA} also increases from 0 to only 0.2% (six false peaks ≥ 0.395). At P_{FA} = 1.08%, P_C = 98.18% (only one missed test FP), but this occurs at a low T = 0.34. To obtain P_C = 100%, a much lower T = 0.245 is needed and this results in a much larger P_{FA} = 12%. One would not operate at such low T levels. The three largest false peaks (false alarm) values are 0.556, 0.473, and 0.425.

The present SDF results show that these filters are useful and achieve good elastic FP distortion recognition and discrimination; P_C = 92.7% with P_{FA} = 0.067% (only 2 errors out of 2970 possible false inputs tested). Better quality FPs are expected to improve P_C, by increasing all minimum true peak values (≈ 82% of the present test set give true correct filter peaks ≥ 0.475, a reasonable value). The true filter correlation peak is the largest one for all but two of the test inputs (B-7, C-5); both of these cases give low true correlation peak values (0.44, 0.25).

7.3.1.3 Average Filter Analysis

Average filter results (Figure 13c) are now discussed. Results are again similar and as expected. Using fine-aligned vs. coarse-aligned data gave large P_C improvements of ≈ 23.5% (from 65.4% to 89.1%) with no false alarms and a 22% improvement in P_C (72.7% to 94.5%) at P_{FA} = 0.067% (two false alarms). fine-aligned data is thus preferable and is now discussed.

With fine-aligned data, P_C ≈ 98% at P_{FA} = 1.2% (but this is a large number of false alarms, 356 out of 2970). The largest false peak value is 0.41 (there are 20 false peaks with values ≥ 0.27). As T varies from 0.41 to 0.33, P_C increases from 89.1% (with no false alarms) to 94.5% at P_{FA} = 0.067% (two false alarms). At T = 0.41, there are no false alarms, six rejected inputs, and P_C = 89% (6 out of 55 misses), but this is a low T. The correct filter output is the largest for 98.2% of the test set (for all but one test input, B-8, which gave a very low true filter test peak of only 0.245) and all false peaks are < 0.41. Better quality FPs are expected to improve results by increasing minimum true filter test peaks. Thus, these simple average filter distortion-invariant filters perform very well. However, at P_C = 96.4%, true peaks are only ≥ 0.27. Average filter FOC data are slightly better than SDF filter ROC data, but the T values used are lower.

The P_C increases using fine vs. coarse-aligned data reflect larger test FP outputs for the correct filters (65.4% above 0.385 for coarse alignment and 89.1% above 0.41 for fine alignment for the average filter, and 51% above 0.55 for coarse-aligned and 78.2% above 0.56 for fine-aligned SDF filters). These *increased true test peaks are due to better rotational alignment of the test input*. As before, we also notice a decrease in the false peak values when fine-aligned data are used. For coarse-aligned SDF data, there are 19 false peaks ≥ 0.395 out of 2970 possible cross-correlations (0.64%) and only six false peaks ≥ 0.395 for fine-aligned data. For average filters, fine-aligned data filters gave 20 false peaks ≥ 0.27 vs. 37 false peaks ≥ 0.27 with coarse-aligned data filters. We attribute the *decrease in false peak values to the better structure present in the filter* (more ridges and valleys are seen and more contrast is expected) when fine-aligned training data is used to form the filters. This gives the filters more discriminating ability. In a later section (Section 7.6.6), we show this by analyzing reconstructed images of coarse and fine-aligned filters. *Thus, all filters benefit from use of fine-aligned data*. A later section will also discuss the quality of the various test (and training images).

7.3.1.4 Comparison of Average and SDF Filters

We now compare SDF and average filter results. Only fine-aligned data filters are considered. In the low P_{FA} region of interest, both perform comparably $P_C = 94.5\%$ with $P_{FA} = 0.067\%$ for the average filter vs. $P_C = 92.7\%$ with the same $P_{FA} = 0.067\%$ for the SDF filter. At the same $P_C = 96.4\%$, the average filter gave $P_{FA} = 0.67\%$ and the SDF filter gave a slightly better $P_{FA} = 0.61\%$. These differences are not significant. Results with a larger test set are needed. Recall that low pass filtering and normalization are used in both cases. This tends to make all filter results more similar. The SDF filter is somewhat preferable as it provides larger true class output peaks. For the SDF filter, $P_C = 94.5\%$ at a larger $T = 0.395$; while for the average filter, $P_C = 94.5\%$ occurs at a lower $T = 0.33$ level. Both true and false peaks are lower for the average filter (no false peaks > 0.41, the second largest false peak is 0.33) vs. the SDF filter (its largest false peak is 0.475, the second largest false peak is 0.43). Whether these trends continue when more test inputs are used should be addressed.

7.3.2 Need for Normalized Data

We next consider the improvement provided by use of normalized data; this is expected to aid the many cases when only a partial FP (such as Figure 4c) input test image is present. None of the test images used were significant partial FPs, thus this issue did not enter our present data. When normalized data are used, the final 350×350 pixel image (Figure 3b) is normalized (by the square root of its energy) before being used in filter synthesis or as a test input. For unnormalized data, this normalization step is omitted. We initially considered use of normalized test data so that partial FPs will have the same input energy as full FPs and are thus expected to give comparable output correlation peaks to those of full FPs. The SDF and Minace filters are designed to give correlation peaks of one for training set FPs after high-pass filtering (by R or c). Normalization is done after high-pass filtering. Normalization of training set data is expected to greatly aid average filters, since partial FPs and all training set FPs will now yield approximately the same output correlation peak value of one. Since the Minace and SDF filter synthesis algorithms produce the same output (one) for all training set FPs even without normalization, they effectively normalize the training set by adjusting the combination weights used (in the Minace and SDF filters); thus, the use of normalized training set data is expected to have less effect on Minace and SDF filters.

We expect filters formed from non-normalized training data to have different weights. If the test input is a partial print and if normalized data is used, then only that percentage of the test print data is considered and the correlation output is not reduced due to the fact that the test input is a partial FP. This initial motivation for use of normalized FP training and test set data (due to the large number of partial FPs, 70% of all FPs have ≈30% of the FP missing) is of valid concern. However, none of the test FPs (only one per finger) used are significant partial FPs (this was by design). Thus, this benefit of normalization will not be seen in our present tests. However, *in general, we expect use of normalized data to aid tests on partial input FPs.*

Use of normalized FPs was found to greatly help other FP test cases, specifically FPs of dry fingers have much white area (higher energy) and low contrast and FPs of oily fingers have much dark area (less energy). The dark ridges in oily prints flow together when the finger is pressed down. Without normalization, dry FPs have high energy and thus give large correlation outputs (for all filters), and oily FPs have low energy due to large dark areas and thus give low true correlation outputs (causing a low T to be needed). Such FP variations must be accepted in practical cases and normalization addresses such cases. *These dry and oily FPs represent the dominant cases in which use of normalized data gives better results for the present database.* Normalization of dry or oily test inputs obviously helps. Normalization of the training set data also produces better filters; although, since all training set data for each person were obtained at the same session, the data for each FP tend to be similar (all oily etc.), thus reducing this benefit of normalization. However, for comparison between filters from oily and dry etc. FPs, normalization greatly helps. All individuals tested were asked to place their finger in an oil mixture before inputting it to the scanner. Most did not, however some had to or their FPs would be too dry to use.

In future filter synthesis, all training set images would not automatically be included in the filters, as was done in our initial filter synthesis. In such cases, use of normalized FP training set data is of more concern, since it will allow selection of training set FPs to use based upon the structure or pattern of the FP rather than based upon the energy of the FP.

Figure 14 shows ROC results for verification tests for the three filter cases when fine alignment was used with and without use of normalized training and test data. Use of normalized data is seen to be better than use of unnormalized data for all filters. Note that the P_{FA} scale for the Minace filter case (Figure 14a) is very different from those of the other filters. For the Minace filter, at a low $P_{FA} = 0.03\%$ (one error), use of normalized data improves P_C by 5.5%. For the SDF filter, the improvement in P_C is ≈7% at $P_{FA} = 0.03\%$. For the average filter, P_C improves by 5% at $P_{FA} = 0.17\%$ (5 errors). Thus, use of normalized data is clearly recommended. Use of normalized data generally increases the lowest true peak (from 0.52 to 0.56 for the Minace filter, and from 0.22 to 0.24 for the average filter, for the SDF it drops from 0.28 to 0.25). Use of normalized data also reduces the maximum false peak (from 0.67 to 0.48 for the Minace filter, from 0.65 to 0.56 for the SDF filter, and from 0.45 to 0.41 for the average filter). These lower false maxima allow use of lower T with better P_C and P_{FA} results. This is the origin of the present better ROC results with normalized data.

We will analyze the specific errors, and FPs and filters that are often problematic and why in a later section (Section 7.6).

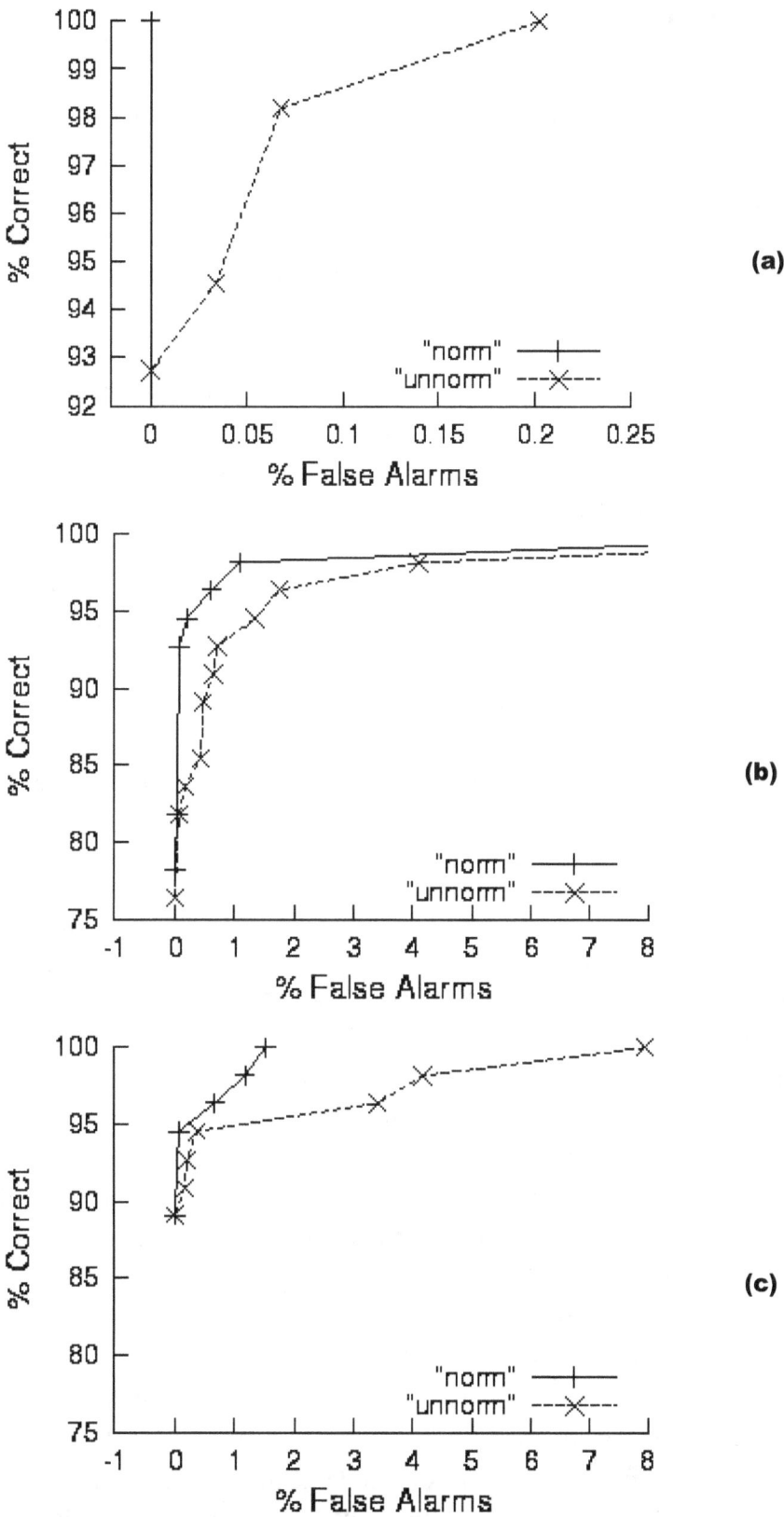

Figure 14. Normalized vs. unnormalized improvement (with fine alignment). ROC verification results for fine shifted data for the Minace (a), SDF (b), and average (c) filters.

7.4 Verification Results

In Section 7.3, we determined that future tests would use fine-aligned and normalized data for all filters. Thus only fine-aligned and normalized data results are considered here.

In verification tests, the user inputs his FP and he states his (or her) identity. In evaluating filters for use in verification, each FP test input is still compared against all 55 filters. Any non-correct filter output above a threshold T is scored as a false alarm. As the threshold T is reduced, more correct filter outputs are greater than or equal to T (performance approaches P_C = 100% correct recognition); but a high percentage of false alarms P_{FA} can now occur. The same threshold T is applied to all filter outputs for all test FPs; thus, a given test input can have its correct filter output largest but, if the correct filter output for some other test inputs are low (above some T), use of this T can produce false alarms even for the test input whose correct filter is largest. There are a maximum of 55^2-55 = 2970 false alarms, P_{FA} is given as a percentage of 2970; P_C is a percentage of 55; test inputs with the correct filter output < T contribute to rejects P_R (a maximum of 55). In practice T is set at a given P_{FA} rate, such as 0.1% (more test image FP sets than we presently have are needed to obtain such data). For a given input, if the correct filter gives an output below T, that test FP is rejected (the reject rate P_R is 1-P_C for verification evaluation. In this reject case, the user is asked to reenter his FP. In our tests, only one test FP is used for each FP; in a final system, the user could enter several test dabs and the system could accept the user if the proper filter output exceeded T for some percentage of the test dabs.

Figure 15 shows ROC (P_C vs. P_{FA}) verification data for the three filter for the low P_{FA} portion of interest. fine-aligned and normalized data are used. We also include results for a one-to-one filter in Figure 15; in this last case, the filter is simply one of the training images with high-pass filtering applied to the training image tested vs. all test inputs. The one-to-one filter performs worst as expected (unless the elastic distortion in the test input is similar to that in the single training image chosen, a low T will be needed and many false alarms are expected). The SDF and the average filter perform comparably. The Minace filter performs perfect (P_C = 100% with no false alarms P_{FA} = 0%). Thus, the combination weights used and the high-pass filtering performed by the Minace filter are preferable to those used in the other filters and our simple high-pass filtering performed. The results for the other filters are expected to be much worse if high-pass filtering was not done (it reduces false alarms as seen in Figure 10).

For these verification tests, the Minace filter performs best (with perfect P_C = 100%, P_{FA} = 0% and P_R = 0%). All other filters considered performed worse, with 21.8% misses (P_C is only 78.2% at P_{FA} = 0) for the SDF filter and with slightly better P_C = 89.1% for the average filter at P_{FA} = 0. As T is lowered, P_C improves rapidly and P_{FA} becomes worse. For the SDF data, T varied from 0.55 (with P_{FA} = 0 and P_C = 78.2%) to T = 0.36 (a very low value) with P_C = 96.4% and P_{FA} = 0.61%). The average filter data is slightly better with P_C = 94.5% with a small P_{FA} = 0.07% (2 errors) at a very low T = 0.33 value. As P_{FA} increases, the difference in P_C performance between the SDF and average filters decreases (but P_{FA} is too large to be of use). Results for P_{FA} > 0.067% for the average filter and for P_{FA} > 0.61% for the SDF filter are not too meaningful since T $\leq$ 0.36 for the SDF filter and is lower (T $\leq$ 0.33) for the average filter for these P_{FA} ranges. Since we

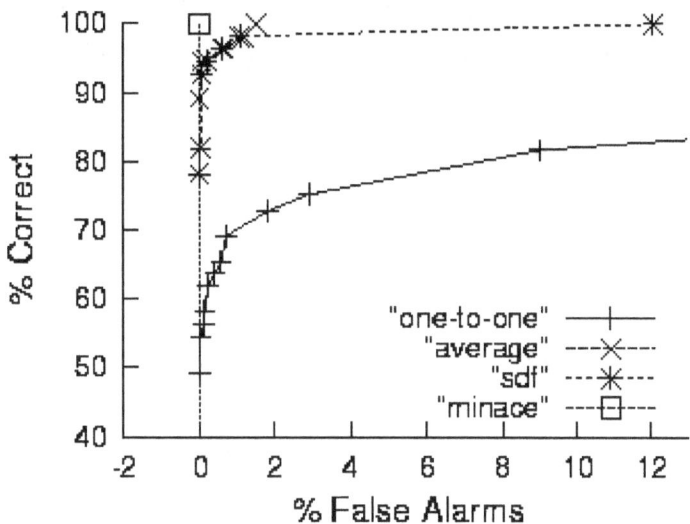

Figure 15. Verification test results (fn data).

would never use such low T values, data beyond these low P_{FA} values are of use only in comparisons of the performance of different filters.

The problem is that several of the filters do not perform well (they do not give high outputs for their single test inputs): four of the SDF filters (7.3%) give outputs < 0.43 for their test FPs, ten of the SDF filters (21.8%) give outputs < 0.475 for their test FPs; the largest false peak is 0.56. Six of the average filters (10.9%) give outputs < 0.41 (the largest false peak value) for their test FPs. Two of the SDF filters do not give true outputs above 0.34 for their single test FPs used. This last remark is relevant, since in SDF identification tests with fine-aligned and normalized data (Figure 16b) the correct filter output was largest (perfect P_C) for all but one (1.8%) of the 55 test FPs. From an analysis of these SDF errors (and analogous average filter errors), filter improvements can be expected and the P_R reject rate can be reduced. This is addressed in a later section (Section 7.6).

At low P_{FA}, both the average and SDF filters give similar performance in Figure 15, and both perform worse than the Minace filter. High-pass filtering was applied to both the average and SDF filter training set data (thus they are not the conventional average and SDF filters), and effectively applied to their test set inputs (by the filtering), all training images were included in both filters, and normalization was done after high-pass filtering; these may explain the comparable performance of average and SDF filters.

7.5 Identification Tests

In identification tests, the user does not state his identity. In use for identification, each input test FP is correlated with all 55 filters; if any filter output is $\geq T$, the input is accepted. In our evaluation tests, scoring is similar; only the filter with the largest output greater than or equal to T is accepted. The purpose of the system is to identify the input test FP as a member of the database. If the filter with the largest output is correct, it counts toward P_C, otherwise it counts toward P_{FA}. P_C is the percent of all 55 test inputs correctly recognized. There are thus a maximum of 55 false alarms and P_{FA} is a percentage of 55. If no filter output is $\geq T$ for a given test input, the input is rejected and contributes to P_R. For evaluation scores of identification filters, $P_C+P_{FA}+P_R = 100\%$. The same T must now be used for all test inputs.

For certain filters or persons with known poor FP and hence poor filters, one might be able to allow lower T for some filters; but this seems dangerous if the T is too low and if that filter gives high outputs (above T) for other FPs of the same class. *Future work can address this alternate scoring method.*

The only way to achieve $P_C = 100\%$ is if $P_{FA} = 0\%$. Thus, for identificaiton evaluation data, we expect larger P_{FA} values than for verification tests (especially since P_{FA} is out of 55 vs. 2970 possible false test inputs).

Figure 16 shows identification test results for various data preprocessing cases for our three filters. *For the Minace filter (Figure 16a), all data preprocessing choices give the same perfect performance*: $P_C = 100\%$, $P_{FA} = 0\%$ and $P_R = 0\%$. *The correct Mince filter output is always the largest. This is very different from other cases.* This filter again performs better than (or comaprable to) all others in all data preprocessing cases. The correct Minace filter output is largest for all cases (and the largest false peak is below the minimum true peak for all $55^2 = 3025$ filter cases). This occurs for all fine/coarse alignment and normalized/unnormalized data cases and for all partial/dry/oily/elastic-distorted test FP input cases. This shows the spectacular robustness of the Minace filter. Use of different T values for filters associated with known poor FPs could extend use of this filter even further as noted earlier. These results are very attractive, since, *using Minace filters, we can use coarse-aligned data (this can be automated),* without the need for the manual fine-aligned data case. Since only the largest output from all 55 filters is considered for each test input, no added filter errors can occur in identification if a lower T is used. In this sense, verification is more demanding than identification, since a filter or FP requiring a low T can increase P_{FA} in verification, even if a filter's correct output is the largest.

Figures 16b and 16c show SDF and average filter identification data. Both show the same improvement trends expected (fine/normalized is best, followed by fine/unnormalized, coarse/normalized and coarse/unnormalized data).

We now consider SDF filter performance (Figure 16b). We note that SDFs using normalized vs. unnormalized fine-aligned data have a much better identification performance compared to earlier verification results ($P_C = 98.2\%$ vs. $P_C = 78.2\%$ with $P_{FA} = 0\%$). Thus, SDF filters perform well. For 98.2% of the test set (54 out of 55 test images), the largest SDF filter output is the correct one and there are no false alarms (with our identification scoring used); but a low T = 0.315 was needed. In practice, test prints with a low filter output would be rejected, as the confidence of such classifications is low.

We now disucss average filter data (Figure 16c). It performs well (a bit worse than the SDF filter). At $P_{FA} = 0\%$, it yields a higher $P_C = 98.4\%$ identification rate than its $P_C = 89.1\%$ verification rate. Thus, again, *identification performance is better than verification performance for average filters*. The problem is that this filter yields low true correlation peak values, compared to the SDF and Minace filters. At a very low T = 0.26 value, the correct filter output is largest for 96.4% of the test inputs (two are rejected, they are not errors). At a lower T = 0.245 level, the correct filter is largest for 98.2% of the test inputs. Our concern is that using such low T levels

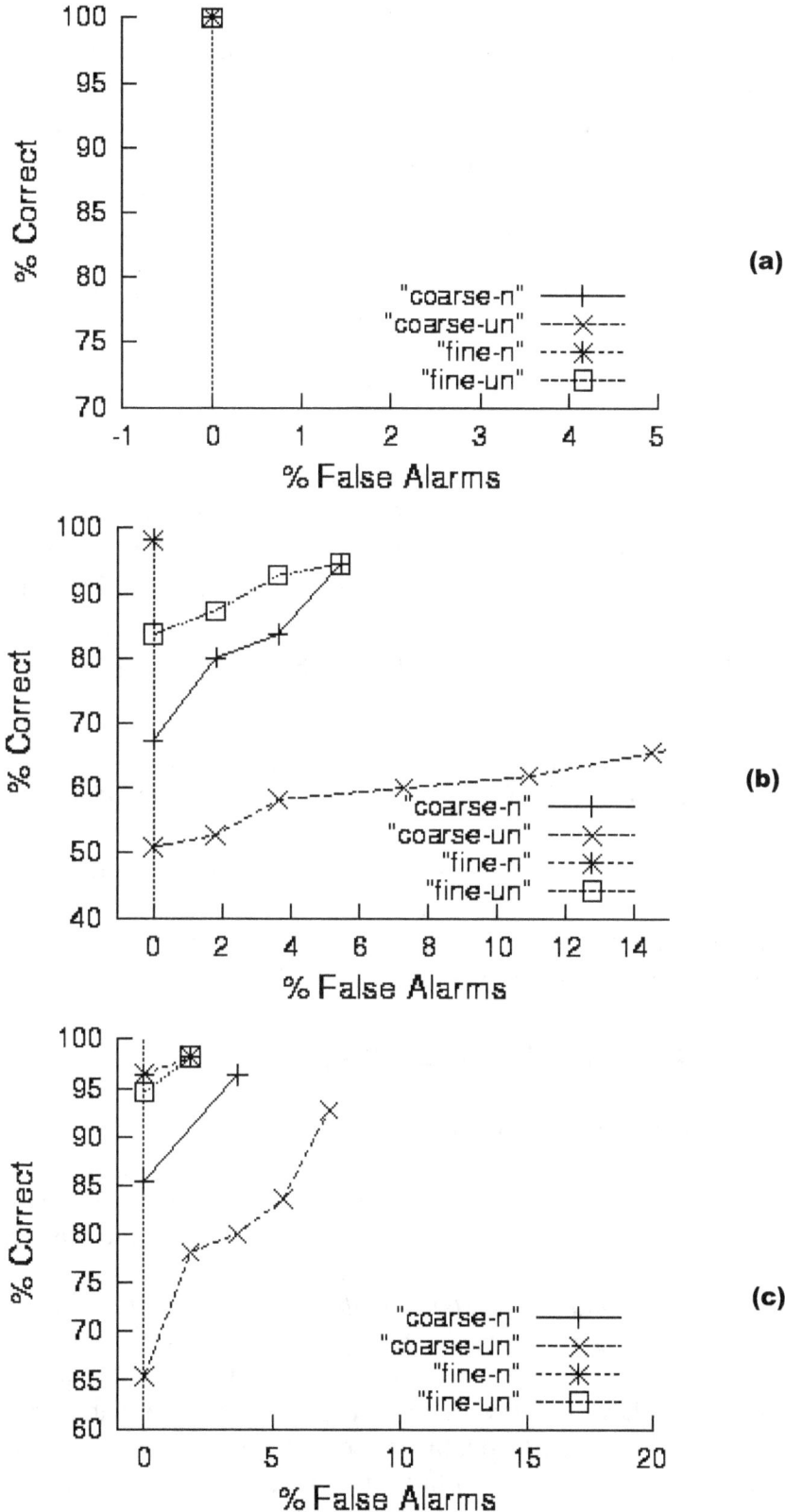

Figure 16. Identification tests for three filters using various preprocessing methods: (a) Minace, (b) SDF, and (c) average filters.

seems dangerous (it will yield P_{FA} errors when more false FP inptus are used in tests). At P_C = 96.4% for identification, $P_{FA} = 0\%$; $P_C = 89.1\%$ (much lower) for verification performance at $P_{FA} = 0\%$. At a very low T = 0.26 level (this seems too low for a realistic system), the correct filter output is still largest for 96.4% of the 55 test cases (only 2 rejects, and no FAs). This property and issue should be addressed in future work ala whether such low T levels should be considered. This requires a larger test set than one FP per person's finger. A larger training and test set will address whether the SDF or average filter is better. Clearly, the Minace filter is best at present.

Most errors in the present identification evaluation tests will be easily handled in practice by rejecting the input FP and requiring the person to reinsert his FP data into the system. This is the present system approach considered for INS ID smart cards. For diverse identification applications, one can envision lowering T for various applications.

Figure 17 shows test results for identification. Table 4 lists low threshold scores. Only fine-aligned and normalized results are included (the best scores for each filter case). The Minace filter performs perfect. The one-to-one filter again performs worst. It correctly identifies only 69% of the test FPs with no false alarms with T = 0.3. As T is reduced, P_C increases and so does P_{FA}. The SDF filter's performance is slightly better than the average filter. $P_C = 98.2\%$ (for the SDF) vs. $P_C = 96.4\%$ (for the average filter) with $P_{FA} = 0\%$ at a low T = 0.315 and 0.26 respectively and $P_C = 98.2\%$ at $P_{FA} = 1.8\%$ for both filters with T = 0.313 for the SDF filter and T = 0.25 for the average filter.

In our identification tests, only one decision is made (based on the largest filter output) for each test input (only one test input is available for each FP finger). With only 55 test input FPs (one per finger), a $P_{FA} = 1.8\%$ corresponds to only one false alarm (FA) and a $P_C = 98.2\%$. As Figure 17 and Table 4 show, the SDF filter gives only one false alarm (FA) at a T = 0.313 and the average filter yields only one FA and one rejected FP at T = 0.256. These FP errors appear to be due to poor quality input test prints, discussed in the next section 7.6. These FPs would typically be rejected and thus these filters could also be useful for FP identification.

7.6 Analysis of Results and FP Image Examples

7.6.1 Database Issues (FP variations present)

Elastic distortions were (and are) the primary distortions to be addressed in our database tests. However, analysis of input FP errors and filters with consistently low values indicated that various other FP variations were also present in the test set data. *Such image variations are expected to arise in reasonable input test data and must be handled by our recognition filters*. These FP variations are shown by example in Figure 18. All FPs shown are test set FPs, that are (ideally) the best FP for a given person's finger. Figure 18a shows a good FP (F-0). It occupies the full 350 pixel diameter. Figure 18b shows an FP with a noticeably higher ridge spatial frequency than those of other people in our database (A-1); this person was the only female in the database and her smaller fingers seem to be the reason for this difference. Her P_C test errors seem to be due to the need for more accurate rotational alignment, in some cases better than our fine rotational alignment, due to the higher spatial frequencies and closer ridge spacings in her FPs. Figure 18c shows an oily test input FP (B-4); as seen it is very dark, has less energy (requiring normalization

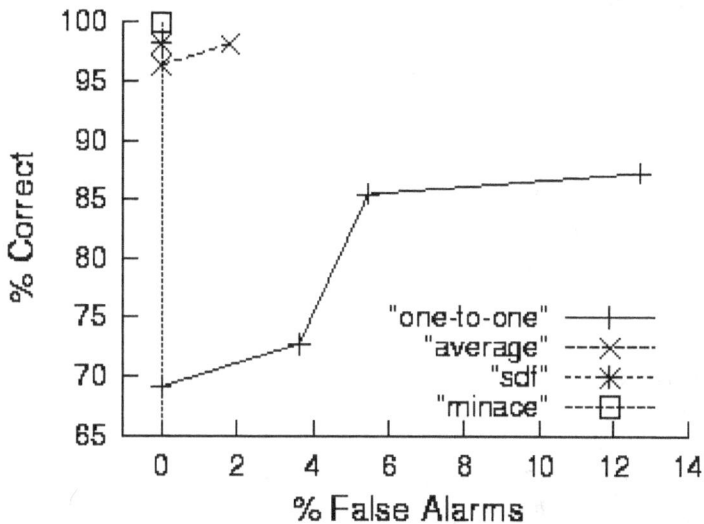

Figure 17. Identification test results.

Table 4: Fine-aligned normalized filter identification performance for different thresholds.

Filter	$P_C(\%)$ (misses)	$P_{FA}(\%)$ (number)	$P_R(\%)$ (number)	Threshold
Minace	100	0.0	0.0	0.48-0.56
SDF	98.2(1) 98.2(1)	0 1.8(1)	1.8(1) 0	0.315 0.313
Average	96.4(2) 98.2(2)	0 1.8(1)	3.6(2) 1.8(1)	0.26 0.256

for good performance) and the ridges and valleys are less clear and are broken-up and have lower contrast. Figure 18d shows a dry test input FP (C-0); it is whiter, has more energy, the ridges are broken-up and less clear and contrast is lower (since dry FPs do not make good contact with the scanner). Figure 18e shows a partial test FP (G-O). Figure 18b is also a partial FP; normalized data is expected to help such FPs. Figure 18f is a test FP that has scars (C-3). Automated minutae location algorithms can have problems with scarred, dry and oily FPs and may reject such FPs as not being of sufficient quality. Partial FPs may not contain enough area to allow a sufficient number of minutae to be located. Our filters are less bothered by such FP quality factors.

These FP examples in Figure 18 are not isolated cases, many of the test (and training) set images have such quality issues. Thus, an FP recognition system must be able to handle such variations in addition to elastic distortions. With fine-aligned and normalized data, all 3 filters were able to successfully perform verification and identification of all FP variations in Figure 18.

7.6.2 Oily Test Input Effects

Figure 19a shows an oily partial test FP (B-7). Analysis of the performance of the different filter cases for this test input demonstrate many of the expected trends. The average filter with fine-aligned and normalized data gave a reasonable 0.49 correlation peak for the correct filter, no other filter output was above 0.18 for this test input. With fine-aligned but unnormalized data, we expect a lower output since the input test print is dark and has low energy; in this case, the correct filter had a lower 0.37 correlation peak as expected; this was the largest output among all 55 filters, even though it is low (thus no false alarms occur for this test input in identification; in verification, other test inputs have false filters with outputs above 0.37 and thus use of such a low threshold will yield verification errors). In practice, such a low 0.37 threshold would not be used. Performance using normalized but coarse-aligned data gave an even lower true filter output of 0.30, but it was again the largest output for any filter.

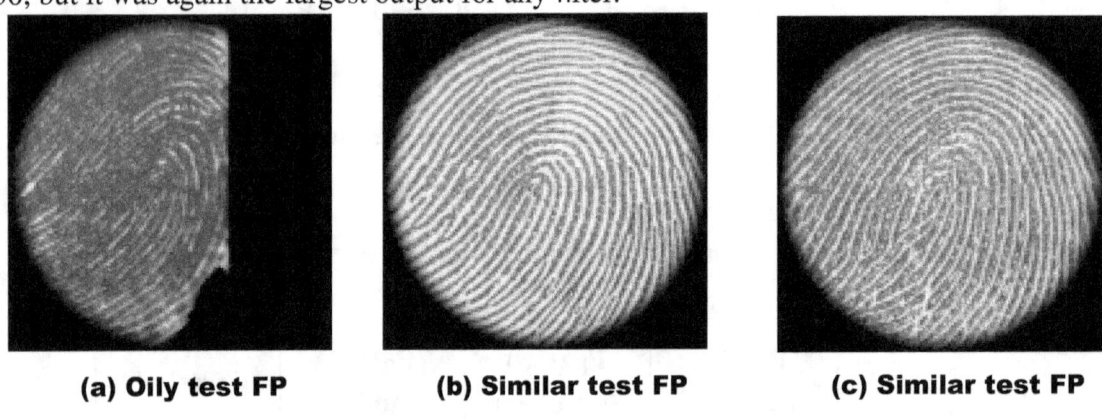

(a) Oily test FP　　**(b) Similar test FP**　　**(c) Similar test FP**

Figure 19. (a) Oily partial test FP and (b,c) two similar type test FPs.

For this Figure 19a test input, the Minace filter with fine-aligned and normalized data gave a large 0.64 correlation peak for the correct filter; no other filter output was above 0.34 for this test input. Using fine-aligned but unnormalized data, we expect a lower correct filter output because the test input is dark and has low energy. The correct filter gave a lower correlation peak value of 0.52, which is quite large, however two other filters (F-5 and C-5) gave larger correlation peaks of 0.67 and 0.56. Both of these other FPs (Figures 19b and 19c) are similar in class (loops) to the test input (Figure 19a). With normalized but only coarse-aligned data, the true filter gave a still lower output correlation peak of 0.44, but it was now the largest of all filter outputs (it would yield no error in identification, but use of such a low threshold would produce errors in verification tests).

7.6.2.1 Further Oily FP Effects

Not all oily prints (like Figure 19a) are problems, especially when the training set also contains oily prints and hence captures the oily nature expected of an input test FP. This occurs when training and test sets are obtained in the same short session; it may not occur in practice when the test sample is entered much later and when FP preparation is required (e.g. to avoid dry FPs). Consider the oily FP In Figure 20 (B-9). Using fine-aligned both unnormalized and normalized data, all three filter types perform well. The correct class filter gives the largest value for all three filters and all true-class filter outputs are large. For the Minace filter, the correct filter output is 0.97 and 1.1 for unnormalized and normalized data. For the average filter, the correct filter output

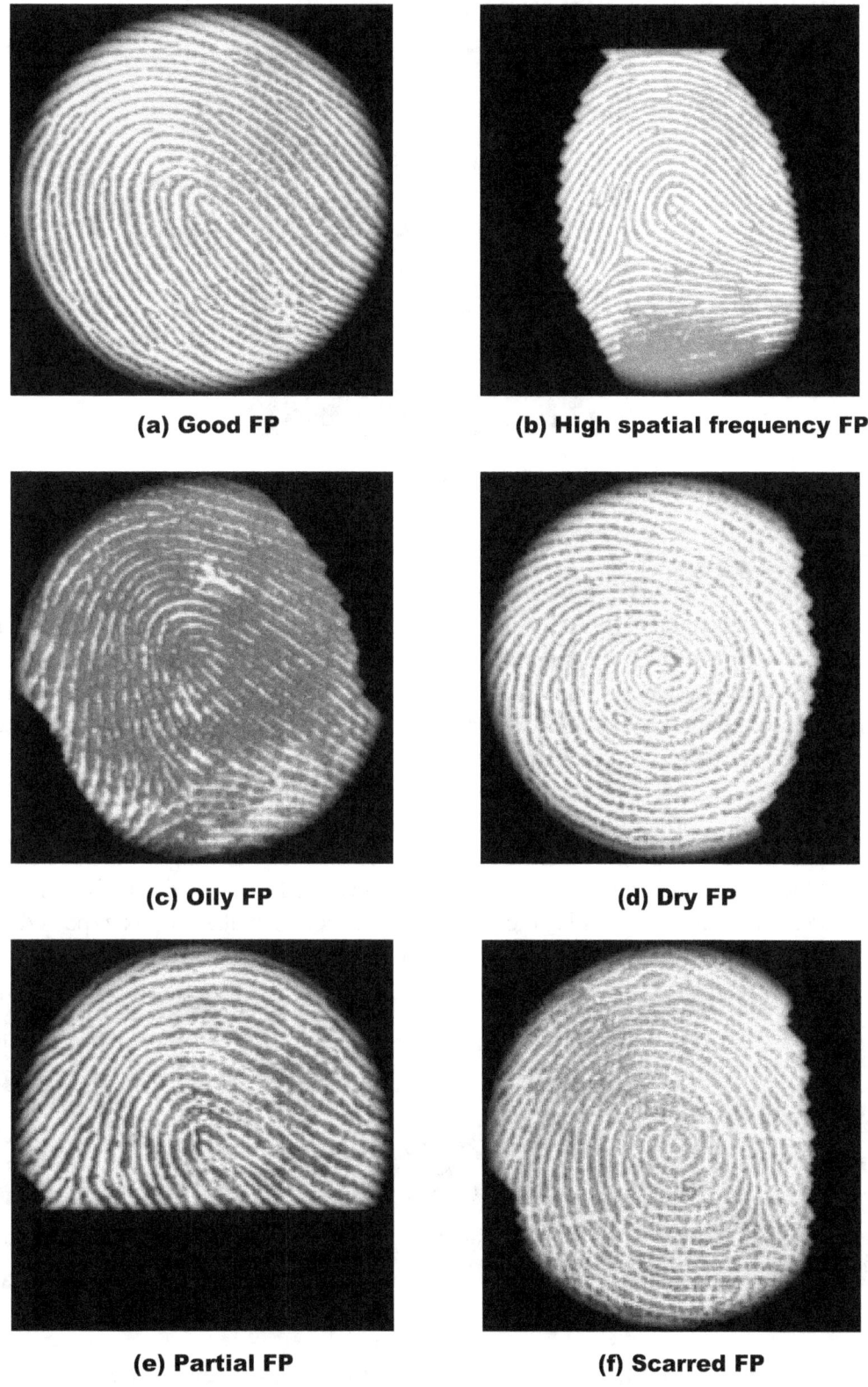

Figure 18. Representative examples of different test FP variations to be expected.

is 0.79 and 0.9. For the SDF filter, the correct filter output is 0.95 and 0.99.

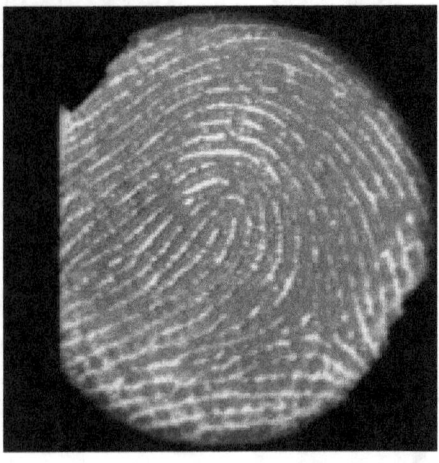

Figure 20. Oily test FP (B-9)

7.6.3 Partial FP Effects

Figure 21 shows a partial test FP (G-0) in Figure 21 and several of the training images (Figures 21 and b) used to form its filter. All training set FPs were partials. This would seem to be a significant problem. However, all filters performed well on this test input. With fine-aligned and either normalized or unnormalized data, the Minace filter gave a large correlation output of 0.78 and 0.84 for the correct filter for this partial test input in Figure 21. The average filter gave large correlation peaks of 0.80 and 0.84 for the correct filter. The SDF filter gave large correlation peaks of 0.84 and 0.96 for the correct filter. Since all training and test images were taken at the same 10 sec session, they are all somewhat similar in the sense that all are similar partial FPs. Thus, the filter captured the partial nature of the test FP and was designed to give peaks of one for partial training set FPs. When the test print is entered at a later time, normalization is expected to help more than it did here and is recommended general practice.

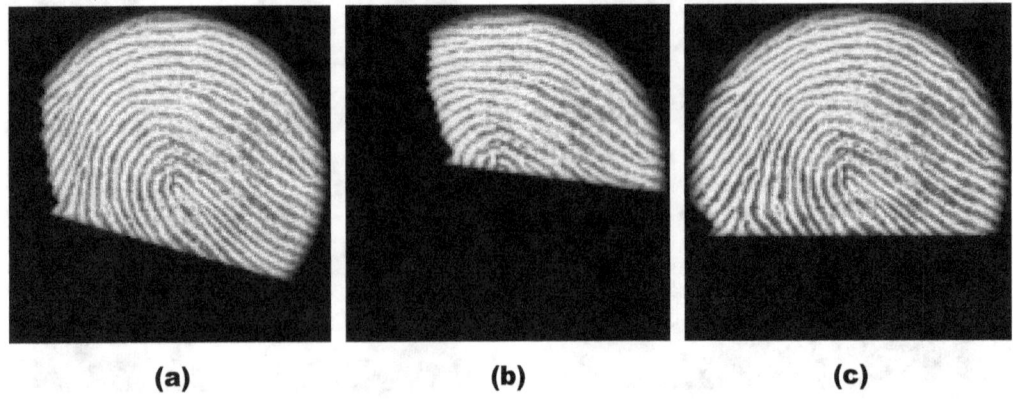

Figure 21. Typical training set (a-b) and test set (c) partial fingerprint example.

7.6.4 Analysis of Lowest-True and Largest-False Test Cases

Analysis of the test inputs for different filters and preprocessing cases was conducted to determine which test inputs gave the lowest correct correlation peak output (the lowest true peak), which filters gave the largest false correlation peak (the largest false peak), whether any test inputs or any filters gave consistent problems (low true peaks or many false peaks). The reason for such occurrences (a poor quality test input FP, a poor quality training set, poor alignment of the trainingn set, or that the test FP elastic distortions did not match those in the training set) were then addressed. To provide such discussion answers, we analzye the test input FP in terms of its quality (oily, dry, partial, scarred, etc.), the training set (is it of poorer quality than the test FP image? etc.), and reconstructions of filters for different cases (the reconstructed image of a filter is expected to be blurred over parts of the FP if alignment etc. is poor). We note at the outset that we found analysis of filter reconstructed images to not provide a clear conclusion and that *this entire filter reconstruction issue merits further attention*. We also note that many poor quality test inputs (e.g. Figures 20 and 21) are not always problematic (for reasons noted). Thus, analyses and conclusions of such marginal test inputs or filters are not *always* true, but they can occur.

For comparison of reconstructed images of filters, Figure 22 shows several examples. Figure 22a shows the reconstruction of a good filter (F-7). It is useful for comparsion; it has significant structure. Figure 22b shows the reconstruction image of a filter (C-5) formed from training set images that are not well aligned (it is blurred and has little structure in portions of the image), especially in the center of the image. Figure 22c shows the reconstructed image (B-5) of a filter formed from oily (dark) training set images. It is quite dark and devoid of detail. All reconstructions are for average filters, since they equally weight all training set images

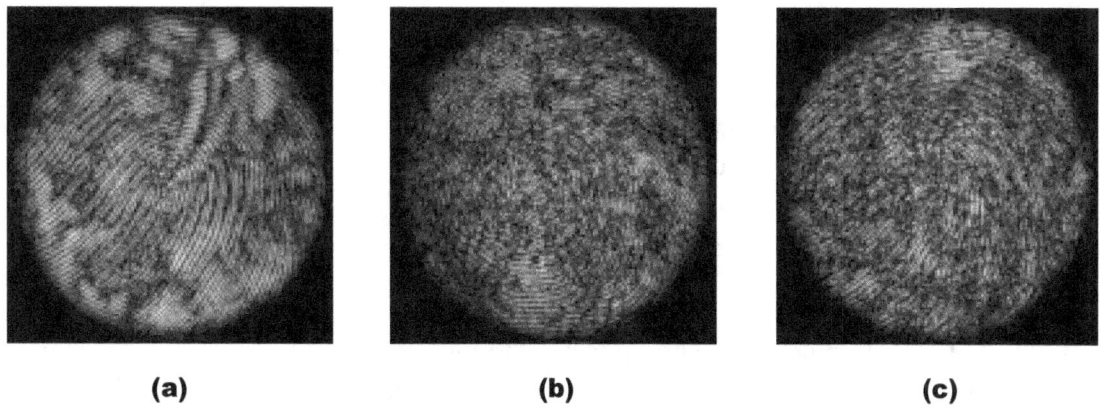

(a)   **(b)**   **(c)**

Figure 22. Reconstructed images of average filters formed from (a) good well-aligned training set images (F-7), (b) not well-aligned training set images (C-5), and (c) poor quality (oily) training set images (B-5).

7.6.4.1 Minace Filter Marginal Case Analysis

For the Minace filter (fine-aligned and normalized data), the minimum true correlation peak was 0.56 (for the B-8 test input); this is a rather large peak value. This test FP is oily and a slightly partial FP (Figure 23a). Its oily nature is expected to be the reason for its low output, plus the associated poor quality of the filter formed from a training set of such FP images. Figure 23b shows the reconstruction of the average filter for this person's finger. As seen, it is very dark and shows little detail (since it is the sum of a number of oily, dark, poor contrast training set images).

The low true correlation peak value for such test inputs could also be due to an elastic distortion difference between the test input and its training set. All of these factors contribute to this lowest Minace true correlation peak; we note that it is a large peak value (0.56) and that no verification or identification errors occur with this Minace filter; thus, this is simply the marginal test set print case. The next smallest true correlation peak (0.58) occurred for the mik-3 input; this FP, Figure 23c, is oily and contains scars; its training set (several training set examples are shown in Figures 23 d-f) contains partial FPs and FPs with scars; these seem to contribute to the lack of detail in the bottom of the FP Minace filter reconstruction in Figure 23g. The largest false correlation peak (a large 0.48) value) for the test input FP in Figure 23a occurs for the F-1 filter; both this (the F-1) FP (Figure 23h) and the test input (Figure 23a) are of a similar loop class.

7.6.4.2 Analysis of marginal FP cases (Average and SDF Filters)

For the average filters using fine-aligned and normalized data, the minimum true correlation peak value occurred for the B-8 FP. The peak value was very low (0.24). The test input FP for this case (Figure 23a) is oily and the training set images used to form the filter are of similar poor quality. In practice, a new training set of images would have been obtained for this case. Two test FPs (E-6 and E-8) from the same person gave low true correlation peak values of 0.45 and 0.46. These test FPs (Figures 24a and b) were somewhat oily (all 10 of these subject's FPs are oily), but are of good quality. These same sta test FPs (Figures 24a and b) also yielded low true correlation peak values of 0.34 and 0.46 for the SDF filters. For one of these FPs (E-8), the training set images (see Figures 24c and d) are worse (oily and scarred) than the test set image and the reconstructed images of its average and SDF filters (Figures 24e and f) are very dark and poor. They lack structure around the center of the image. Thus, in this case, the training set is of poor quality and this accounts for the low average and SDF filter outputs. We note that the Minace filter for this case did give a large 0.63 correct correlation peak value for this test FP (this was the third smallest true correlation peak for the Miance filter); Minace filter values are consistently larger. A detailed examination of the fine-aligned training set images showed shifts in them (i.e. incorrect alignments). This was observed for both FPs (E-6 and E-8); this could also have affected filter quality. When this occurs, we find that the center of the reconstructed filter image lacks structure (Figures 24e and 24f). It indicates the possible need to modify our fine-alignment algorithm (we used the maximum correlation peak value). The maximum correlation peak value (0.41) from a false filter occurred for the E-7 test input and the F-6 filter. These two FPs were of the same class (loop) and are shown in Figures 24g and 24h.

SDF filter marginal cases for fine-aligned and normalized data are now discussed. The minimum true correlation peak output for this case was very low (0.25) and occurred for the mcc-5 FP. This test FP is of good quality (Figure 25a). From a detailed analysis, *the fine-aligned training set images for this filter (aligned by largest correlation peak) were found to have their FP centers shifted*. This shows up in the filter reconstruction image in Figure 25b; the center of this image lacks structure. This seems to be the reason for its poor performance. However, we note that the average filter and the Minace filter for this C-5 input FP (using the same aligned training set images) performed well (true correlation peak values of 0.73 and 1.06); in all cases, all filters used training set images with the same alignment.. Thus, a different elastic distortion in the C-5

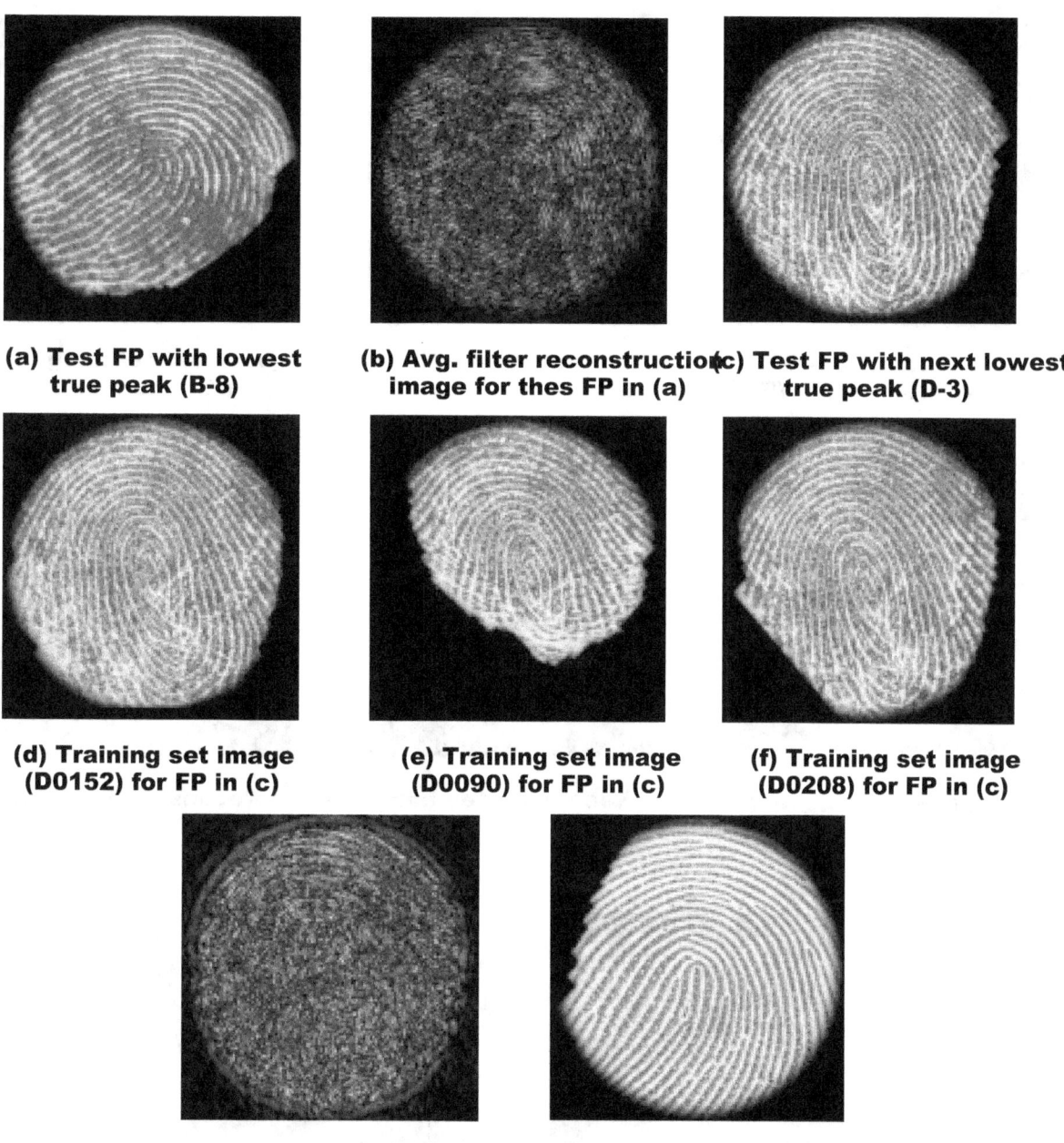

(a) Test FP with lowest true peak (B-8)
(b) Avg. filter reconstruction image for thes FP in (a)
(c) Test FP with next lowest true peak (D-3)
(d) Training set image (D0152) for FP in (c)
(e) Training set image (D0090) for FP in (c)
(f) Training set image (D0208) for FP in (c)
(g) Minace filter reconstruction (fn-D-3) for the FP In (a)
(h) Test input (F-1) for filter with largest false response for (a) input

Figure 23. Minace filter (fn) analysis

test FP from those in the training set does not seem to be the cause of this low true class filter correlation peak (rather the weights applied in combining the different training set images and the high-pass preprocessing are the differences in the different filters). The next lowest true class peak (0.34) is also very low and occurred for the test FP E-6 (Figure 24a); as discussed earlier, improper shifts in fine-alignment of the training set images for this FP (and possibly differences in its elastic distortions) are the reasons for the poor performance of this test FP. Two other low true peak outputs occurred for the B-7 (Figure 19a) and the E-8 (Figure 24b) test FPs (peaks of 0.44 and 0.46). These cases were discussed earlier. The same test and training set remarks apply here

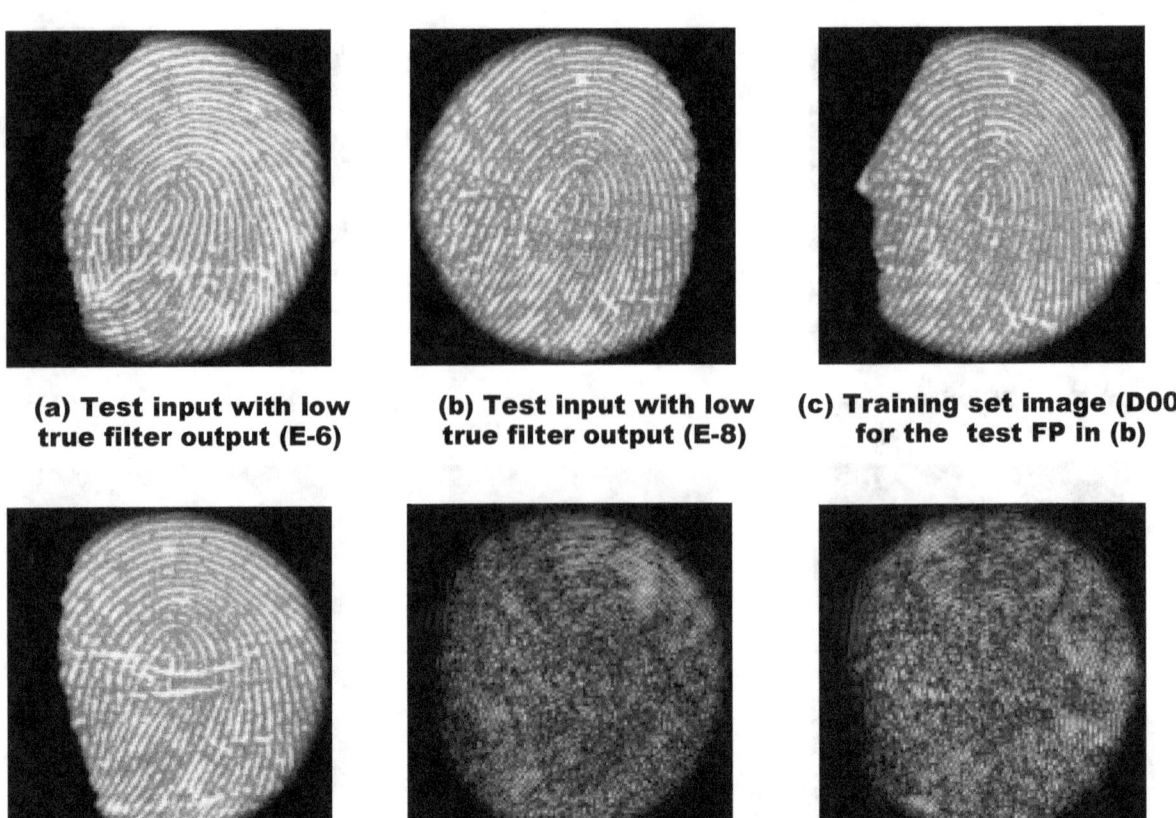

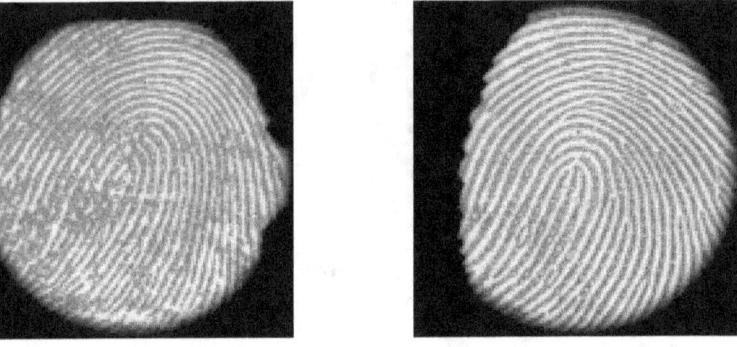

Figure 24. Average filter (fn) analysis.

also to explain these SDF filter remarks.

The largest false class SDF filter peaks are now discussed. The largest false class SDF filter peak (0.56) occurs for the B-7 (Figure 19a) input test FP and the F-6 (Figure 24h) filter. The B-7 test FP (Figure 19a) is very poor and both the B-7 and F-6 FPs are of the same class (loop). The next largest false class peak (0.47) occurred for the F-7 test FP input (Figure 25c) and the E-7 filter (Figure 25d shows its test input). These FPs are also of a similar class (loop) and the test and training (person E) FPs are oily.

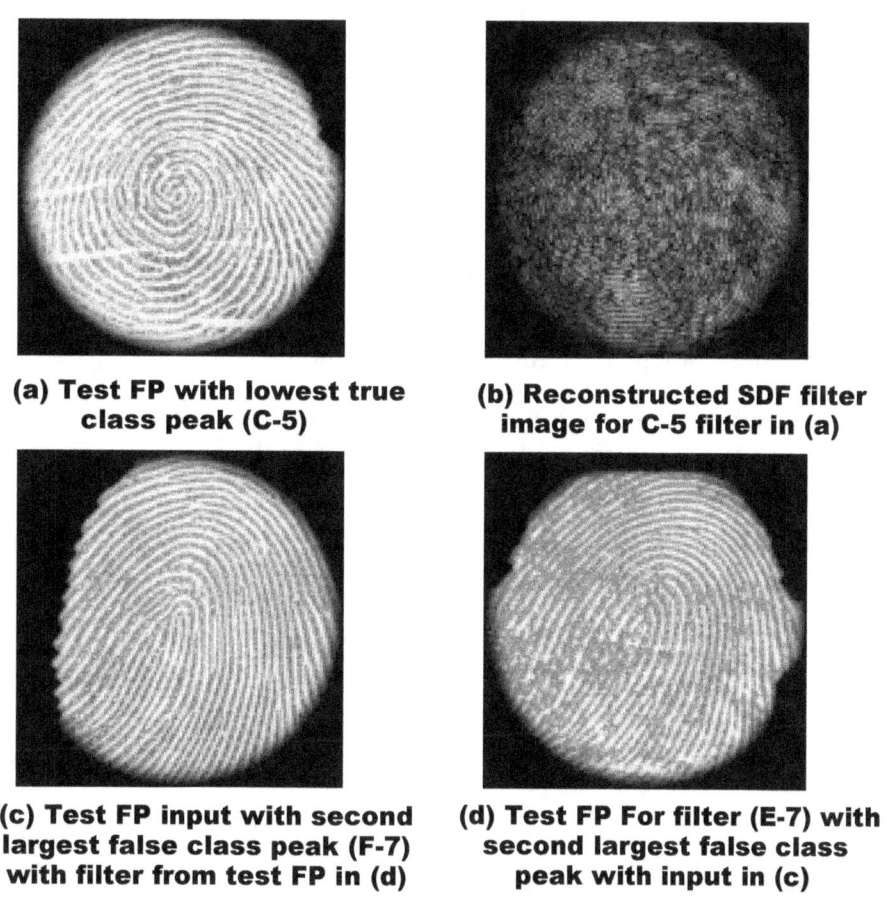

(a) Test FP with lowest true class peak (C-5)

(b) Reconstructed SDF filter image for C-5 filter in (a)

(c) Test FP input with second largest false class peak (F-7) with filter from test FP in (d)

(d) Test FP For filter (E-7) with second largest false class peak with input in (c)

Figure 25: SDF marginal (fn) filter cases.

All marginal cases and false alarms cannot be exactly explained, but the prior remarks provide useful explanations, discussions and guidelines for most cases and are of use in defining future work needed.

7.6.5 Rejection Performance (P_R) Analysis

For verification tests, if no filter gives an output above some threshold (T), then the associated test input is rejected (it contributes to P_R) as being wrong or of poor quality. For this verification case, $P_R = 1-P_C$. For identification tests, only the largest filter output above T is considered. If it is wrong, this is a false alarm and contributes to P_{FA}. If no filter output is above T, that test input contributes to P_R. P_R is a percent or fraction of all 55 test FPs. For this identification case, $P_C+P_R+P_{FA}=1$. *We expect filters to be better able to handle FPs of poor quality that may otherwise be rejected by a minutae matching FP system. In many cases, this is the desired case.* We now briefly analyze this P_R issue for the different filters, considering only filters using fine-aligned and normalized data. In some cases, we consider other filter cases (Minace filters with coarse-aligned normalized data).

Minace filters give perfect verification performance (P_R and P_{FA} are 0) with fine-aligned and normalized data. They give perfect performance for all data preprocessing cases for identification tests. We thus consider results with coarse-aligned normalized data for verification (since coarse alignment can be automated easily). The lowest true peak (0.41) for the kle-8 test FP is too low of a T to use (67 out of 2970 false alarms result). The largest false peak (0.546) occurs for a wat-1 test FP input and the wat-3 filter (poor coarse training set alignment can explain this case). Thus, we set a threshold T = 0.55. At this T, there are no false alarms, but P_C = 74.5% is low (41 of 55 test inputs are correctly classified) and P_R = 25.4% (14 rejects out of 55 test inputs). Thus, Minace filters using coarse-aligned data merit further analysis and study.

At T = 0.41, average filters give no false alarms, P_C = 89% (49 of 55 test FPs correctly recognized) and P_R = 10.9% (6 rejects) for verification tests. If several false alarms are allowed, at T = 0.33 (this is very low) one can achieve P_{FA} = 0.67% (two false alarms) with P_C = 94.5% (52 of 55 correct) and a lower P_R = 5.4% (three rejects). For identification tests, with no false alarms, at T = 0.26 (this is too low), we achieve P_{FA} = 0, P_C = 96.4% (53 of 55 test FPs correct) and P_R = 3.6% (two rejects).

SDF filter performance is similar. At T = 0.56, there are no false alarms (P_{FA} = 0%), but P_C = 78.2% is low (43 of 55 recognized correctly) and there are 12 rejects (P_R = 21.8%). If two false alarms can be tolerated (P_{FA} = 0.07%), we achieve P_C = 92.7% (four misses) and P_R = 7.3% (four rejects).

7.6.6 Use of Fine vs. coarse-aligned Data (Analysis via Reconstructed Filters)

Fine-aligned data is generally expected to produce filters with less blur and with more structure and to thus yield better results. To see this, we show normalized average filter reconstructions for the wat-2 FP with coarse (Figure 26a) and fine (Figure 26b) aligned training set data. As seen and expected, the filter reconstructions have much more structure and detail when formed from fine-aligned data.

Figure 27 shows another example for the F-7 FP average filter.

7.6.7 Normalized Vs. Unnormalized Data Examples

Use of normalized vs. unnormalized data (fine-aligned) improves scoring in most cases. Figure 28 shows examples of various test inputs whose filters exhibited improved results when formed from normalized data. Many other similar examples occur in the database. However, we note that this conclusion is not a general one for all cases. There are many cases where use of normalized

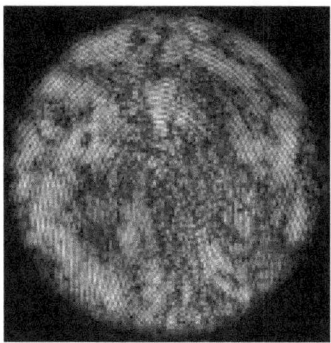
(a) Formed from coarse-aligned data

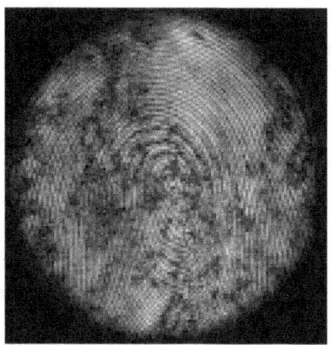
(b) Formed from fine-aligned data

Figure 26. Average filter reconstructions for the F-2 FP using coarse (a) and fine (b) aligned training set data.

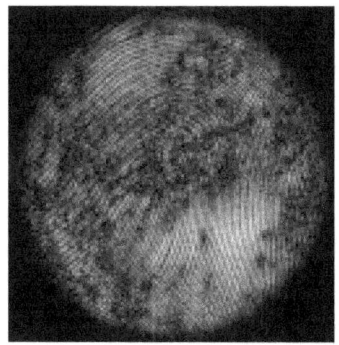
(a) Formed from coarse-aligned data

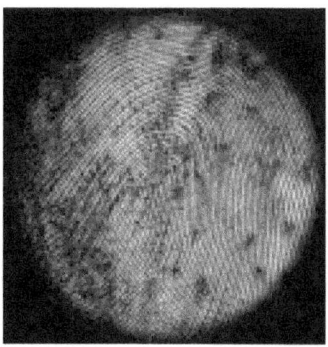
(b) Formed from fine-aligned data

Figure 27. Average filter reconstructions for the F-7 FP using coarse (a) and fine (b) aligned training set data.

data does not improve results, however, the decrease in performance (test FT output) is generally small in all cases that had large correlation peak outputs; conversely improvements using normalized data noticeably improved previously low correlation peak outputs cases so that they exceeded threshold (T) and improved false alarms or missed detections.

Data results are confusing and are included to document them, and to note that we need further tests to understand them. For the test FP (B-7) in Figure 19a, use of normalized vs. nonnormalized data improved the minimum true filter output for the Minace filter (the minimum true test set correlation peak improved from 0.52 to 0.64) and for the average filter (from 0.38 vs. 0.49), while for the SDF filter it decreased for some unknown reason (from 0.58 vs. 0.44 for unnormalized vs. normalized data.

We note that many cases show a noticeable change or a loss in correlation peak value using normalized vs. unnormalized data.

7.6.7.1 Recognition of FPs with Scars

Figure 28 shows two test inputs (D-4 and D-9) that possess scars and that are successfully recognized by all three filter types using fine-aligned and normalized data. Many more similar cases exist in our database. Such FPs occur quite often and many of them would be rejected by minutae matching FP systems as having poor quality. Our filters handle such cases quite well.

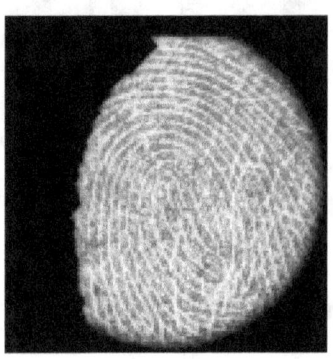
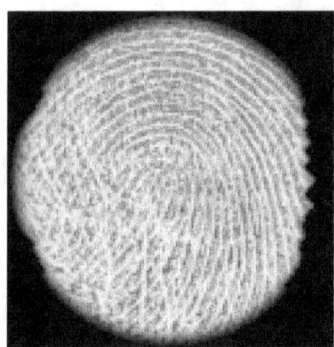

(a) D-4   (b) D-9

Figure 28. Two scarred test FPs successfully recognized by all of our filters.

7.7 Further Tests (Number of training images per FP)

For initial tests, we selected FPs that had at least 9 elastic versions, thus allowing use of at least 8 training set elastic distorted images to be included in the filter for each FP. As noted (Sections 7.3.2 and 7.3.3), our Minace filter gave perfect results for the 55 FPs (out of 200) having at least 9 FPs. We briefly tested the other 145 FPs in the full database of 200 FPs (with less than eight FP versions available). The goal of this advanced test is to determine if fewer than eight elastic distorted FPs is an adequate database for filter synthesis. Only Minace filters were considered with the c value determined in Section 6 used and with all available training set images included in the filters. Figure 29 shows ROC verification data for the case of filters formed with different numbers of training set images. The 8-13 case was that of our original data, it gave perfect results. All other cases gave some errors. Thus, it seems that at least eight training set images are needed for some cases.

When filters with only seven FPs in the filter were included (there were 42 such FPs out of 200), P_C (for these 42 FPs) was only 85.7% (36 FPs were correctly recognized and 6 FPs were rejected, $P_R = 14.3\%$) with one false alarm out of 1722 possible false alarms ($P_{FA} = 0.058\%$) at T = 0.523. At a larger T = 0.525, 36 FPs were correct, six were rejected and there were no false alarms. As T is reduced lower, P_C increases and P_R decreases ($P_C = 100-P_R$). At T = 0.42, 38 FPs were correct, four were rejected and five were misclassified. FPs with fewer training images gave progressively worse results. The 34 FPs with six images in the training set gave $P_C = 91.2\%$ (three rejected FPs) with no false alarms at a low T = 0.46 and $P_C = 94.1\%$ (two rejected FPs) with one false alarm, etc. The minimum threshold T for the 55 filters with ≥ 8 training set FPs was T = 0.56 (for perfect results). If the same minimum T were used for all other filters (with less than eight training set

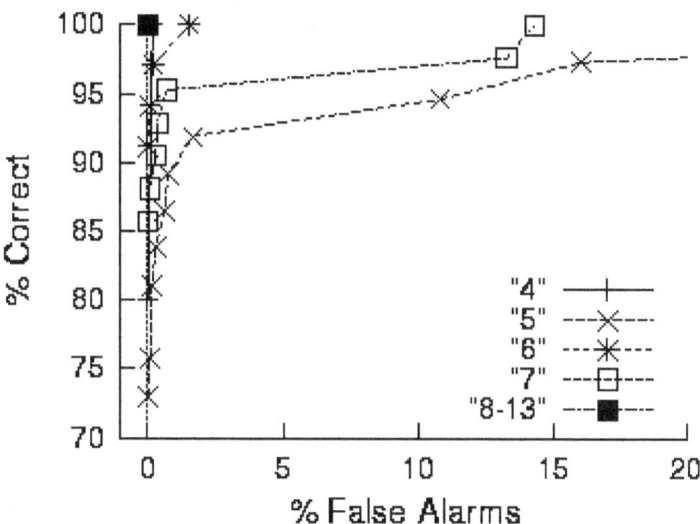

Figure 29. Verification tests on FP Minace filters using only 4-7 training set images per FP.

images), there would be *no false alarms, and 22 rejections* (or about $P_R = 11\%$, for the 193 FPs with at least four training set images). We would not expect to realistically use low T values and we expect to know when a given filter is poor before including it in a final system. Thus, *with realistic thresholds, our distortion-invariant filters do not seem to produce false alarms*. Clearly, at least eight training set images are needed per FP to decrease P_R to low levels.

Identification test results for filters using 4-7 training set images (Figure 30) showed similar trends. Results were better of course, with two of the filters giving perfect $P_C = 100\%$ with $P_{FA} = 0$ (all 25 FPs with four samples, and all 34 FPs with six samples, gave perfect results). Some FPs perform well with fewer FP samples, some give low minimum true or large maximum false correlation peaks (as in earlier data).

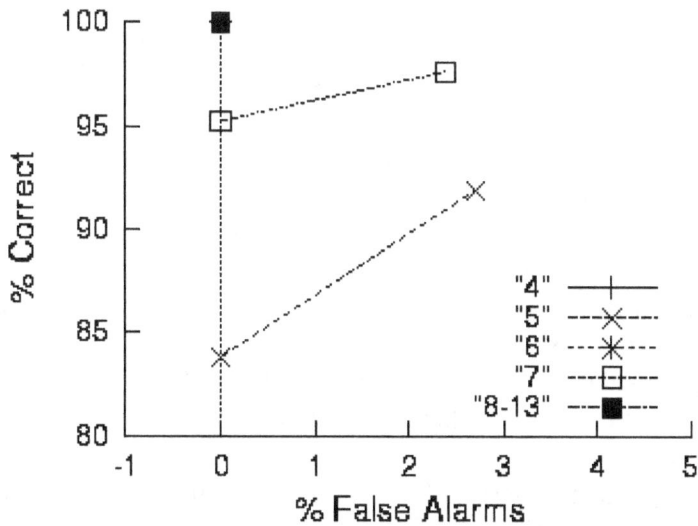

Figure 30. Identification tests on FP Minace filters using only 4-7 training set images per FP.

Figure 31 shows several of the FP test set errors that occurred in verification or ID or both for the Minace filter when only seven rather than eight training set images were present for the given FP. Associated training set images are shown in Figure 32. Figure 31a shows a dry FP with scars present, Figure 31b shows a dry FP with more extreme scars. Such effects plus shifts in the training set images and test set elastic distortions not present in the training set are among the reasons for low true correlation peaks for such test inputs.

7.8 Conclusions

Distortion-invariant Minace filters for live-scan FP recognition with elastic-distortions are very attractive. In initial tests, they provide perfect identification results (even with coarsely-aligned and unnormalized data) and perfect verification results (using fine-aligned and normalized data). As few as eight elastic-distorted images of an FP seem adequate to capture the range of distortions expected. This training set can easily be obtained in a short 10-20 sec training session. Oily, dry, scarred and partial FPs are all handled by this method. We expect that many of these FPs would be rejected by standard minutae-matching systems; distortion-invariant filter FP recognition systems seem to be much more tolerant of such realistic expected FP variations. Such methods merit further research and more extensive testing and analysis. Future work should consider whether a single threshold should be used for all FPs in verification, whether a lower threshold should be used for certain "known" problematic FPs (easily determined during the training stage). It has been noted that 3-4% of FPs are not expected to be easily classified due to such defects and that this percentage may vary notably for certain groups; this issue also merits attention for specific applications.

REFERENCES

1. D. Casasent and C. Wilson, "Optical metrology for industrialization of optical information processing," SPIE, Vol. 3386, pp. 2-13, April 1998, invited paper.
2. K. Karu and A. Jain, "Fingerprint classification", Pat. Recog., vol. 29, pp. 389-404, March 1996.
3. V. Srinivasan and N. Murthy, "Detection of singular points in fingerprint images", Pat. Recog., vol. 25, pp. 139-153, Feb. 1992.
4. N. Ratha, K. Karu and A. Jain, "A real-time matching system for large fingerprint databases", IEEE Trans. PAMI, Vol. 18, no. 8, pp. 799-813, Aug. 1996.
5. A. Stoianov, C. Soutar, and A. Graham, "High-speed fingerprint verification using an optical correlator", SPIE, Vol. 3386, pp. 242-252, 1998.
6. D. Roberge, C. Soutar and B.V.K. Vijaya Kumar, "Optimal correlation filter for fingerprint verification", SPIE, Vol. 3386, pp. 123-133, 1998.
7. J. Rodolfo et al, Applied Optics, 34, 1166, Apr. 1995. NEED TO EDIT AND LOCATE.
8. H. Rajbenbach et al, "Fingerprint database search by optical correlation", SPIE, 2752, pp. 214-223, April 1996.
9. F. Gamble, L.M. Frye, D.R. Grieser, "Real-time fingerprint verification system", Applied Optics, vol. 31, pp. 652-655, Feb. 1992.
10. C. Watson, P. Grother, E. Paek and C. Wilson, "Composite filter for VanderLugt correlator", Proc. SPIE, vol. 3715, pp. 53-59, 1999.
11. B.V.K. Vijaya Kumar, "Tutorial survey of composite filter designs for optical correlators,"

Applied Optics, Vol. 31, 4773-4801, 1992.
12. G. Ravichandran and D. Casasent, "Minimum Noise and Correlation Energy (MINACE) Optical Correlation Filter," Applied Optics, Vol. 31, pp. 1823-1833, 10 April 1992.
13. T.H. Chao, H. Zhou, G. Reyes, "512x512 high-speed gray-scale optical correlator", Proc. SPIE, vol. 4043, pp. 40-45, 2000.
14. H. Zhou, T.H. Chao, G. Reyes, "Practical filter dynamic range compression for gray-scale optical corrleator using bipolar amplitude SLM", Proc. SPIE, vol. 4043, pp. 90-95, 2000.
15. C. Wilson, C. Watson, and E. Paek, "Combined optical and neural network fingerprint matching", Proc. SPIE, Vol. 3073, pp. 373-382, April 1997.
16. D. Casasent and S. Ashizawa, "Synthetic Aperture Radar Detection, Recognition and Clutter Rejection with new Minimum and Correlation Energy filters," Optical Engineering, vol. 36, no. 10, pp. 2729-2736, October 1997.
17. D. Casasent and R. Shenoy, "SAR detection and clutter rejection MINACE filters", Pattern Recognition, Vol. 30, No. 1, pp. 151-161, 1997.
18. E.R. Dougherty, "An Introduction to Morphological Image Processing", SPIE Tutorial Textbook, February 1992, ISBN 0-8194-0845-X.
19. "Optical Hit-Miss Morphological Transform," Applied Optics, Vol. 31, pp. 6255-6263, 10 October 1992 (Casasent, Schaefer, Sturgill).
20. G.T. Candela, P.J. Grother, C.I. Watson, R.A. Wilkinson, C.L. Wilson, "PCASYS - A Pattern-Level Classification Automation System for Fingerprints", NISTIR 5647, August 1995.
21. J.H. Wegstein, "An automated fingerprint identification system", NIST, NBS Special Publication 500-89, February 1982.
23. D. Casasent, "Unified synthetic discriminant function computational formulation", Applied Optics, vol. 23, pp. 1620-1627, 1984.
24. T. Grycewicz, "Fingerprint identification with the joint transform correlator using multiple reference fingerprints", Proc. SPIE, vol. 2490, pp. 249-254, 1995.
25. K. Fielding, J. Horner, C. Makekau, "Optical fingerprint identification by joint transform correlator", Optical Engineering, vol. 30, pp. 195880-jj-1961, Dec. 1991.
26. T. Grycewicz, "Fingerprint recognition using binary nonlinear joint transform correlators", SPIE, Critical Review, CR65, pp. 57-77, August 1997.
27. R.D. Juday, "Optimal realizable filters and the minimum distance principle", Applied Optics, vol. 32, pp. 5100-5111, 1993.
28. D. Maio and D. Maltoni, "Direct gray-scale minutiae detection in fingerprints", IEEE Trans. on Pattern Analysis and Machine Intelligence, vol. 19, no. 1, pp. 27-40, January 1997.
29. A.K. Jain, L. Hong, S. Pankanti and R. Bolle, "An identity-authentication system using fingerprints", Proc. IEEE, vol. 85, no. 9, pp. 1365-1388, September 1997.

APPENDIX A1: NIST Real-Time Optical FP Recognition System

Figure A1 shows the schematic diagram of this frequency plane correlator FP recognition system. Figure 1B shows a photo of it. The input SLM is a Kopin LVGA LC SLM, matched spatial filter (MSF) is recorded on a Newport Research Corp. HC-300 thermoplastic (TP) SLM. The user enters several seconds of data through the live-scan FP reader (Identification Technology) by moving his finger around the input pad to produce five-ten? seconds of various elastic distorted (and rotated) versions of his FP. These training set elastic distorted FP patterns are recorded, stored and fed one by one to the LC SLM; their FTs are summed on the TP SLM to produce a composite filter in the synthesis mode.

After the above training session, the user submits a new test-set version of his print to the reader, it is sent to the input Kopin SLM, correlated with the filter on the TP SLM and if the output correlation peak height at the output plane (CCD-3) is sufficiently high, the test FP input is recognized. Other FPs (false class ones) are also input and tested vs. the filter. Rotational misalignments of the test input print are handled by rotating the test print to achieved the best correlation output. The input SLM is mounted on a rotation stage to achieve this rotational search automatically. No shift search is performed for the input test plane. Rather, a 25×25 pixel region around the center of the output correlation plane is searched for a peak above threshold. An image of the input test print and its FT can be viewed on CCD-1 and CCD-2 respectively. A 10 mW HeNe laser light source was used with a ND2 neutral density filter (why ND used?, saturate CCD output?) (to avoid saturation of the output correlation plane CCD detector used); thus only 0.1 mW of effective laser power is needed. The input light source was pulsed on for ? ms for each test input correlation. The laboratory system can be used to automatically determine the extent and center of each training set input in filter synthesis (this alignment of training set inputs produces better filters).

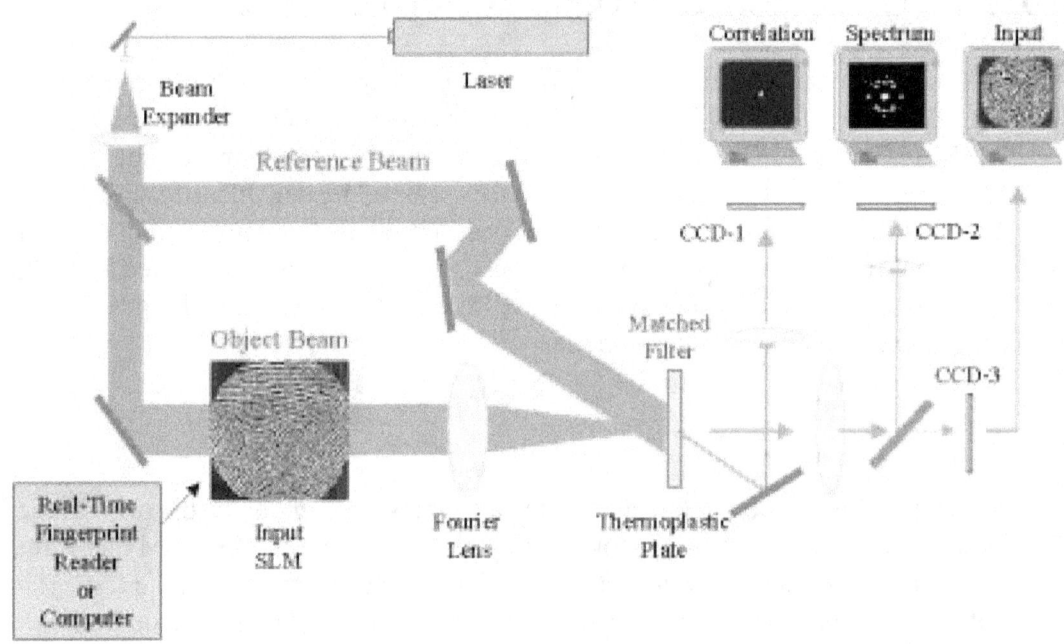

Figure A1. Schematic of the NIST optical FT correlator.

Figure A2. Photograph of the NIST Optical FT correlator.

www.ingramcontent.com/pod-product-compliance
Lightning Source LLC
Chambersburg PA
CBHW081737170526
45167CB00009B/3854